The HUMAN FACTORS of TRANSPORT SIGNS

The HUMAN FACTORS of TRANSPORT SIGNS

Edited by
Cándida Castro
Tim Horberry

CRC Press is an imprint of the
Taylor & Francis Group, an **informa** business

CRC Press
Taylor & Francis Group
6000 Broken Sound Parkway NW, Suite 300
Boca Raton, FL 33487-2742

First issued in paperback 2019

ISBN-13: 978-0-415-31086-4 (hbk)
ISBN-13: 978-0-367-39438-7 (pbk)

Library of Congress Cataloging-in-Publication Data

The human factors of transport signs / edited by Cándida Castro, Tim Horberry.
p. cm.
Includes bibliographical references and index.
ISBN 0-415-31086-5 (alk paper)
1. Traffic signs and signals. 2. Human engineering. I. Castro, Cándida. II. Horberry, Tim.

TE228.H86 2004
625.7'94—dc22 2003068740

Library of Congress Card Number 2003068740

Visit the Taylor & Francis Web site at
http://www.taylorandfrancis.com

and the CRC Press Web site at
http://www.crcpress.com

Table of Contents

Chapter 1
An Introduction to Transport Signs and an Overview of This Book 1
Tim Horberry, Cándida Castro, Francisco Martos, and Patricie Mertova

Chapter 2
History of Traffic Signs 17
Maxwell G. Lay

Chapter 3
Design of Traffic Signs 25
Maxwell G. Lay

Chapter 4
The Effectiveness of Transport Signs 49
Cándida Castro, Tim Horberry, and Francisco Tornay

Chapter 5
Considerations in Evaluation and Design of Roadway Signage from the Perspective of Driver Attentional Allocation 71
Terry C. Lansdown

Chapter 6
Railway Signage 83
Alexander Borodin

Chapter 7
Airport Signing: Movement Area Guidance Signs 95
Kirstie Carrick, Peter Pfister, Robert Potter, and Roy Ng

Chapter 8
The Aging Eye and Transport Signs 115
Donald Kline and Robert Dewar

Chapter 9
Motivational Aspects of Traffic Signs 135
Ray Fuller

Chapter 10
Cross-Cultural Uniformity and Differences in Roadway Signs, Evaluation Techniques, and Liabilities 155
Hashim Al-Madani

Chapter 11
Comprehension of Signs: Driver Demographic and Traffic Safety Characteristics 169
Hashim Al-Madani

Chapter 12
Specific Design Parameters: VMS Part I 185
Luís Montoro, Antonio Lucas, and María T. Blanch

Chapter 13
Some Critical Remarks on a New Traffic System: VMS Part II 199
Antonio Lucas and Luís Montoro

Chapter 14
A Sign of the Future I: Intelligent Transport Systems 213
Michael A. Regan

Chapter 15
A Sign of the Future II: Human Factors 225
Michael A. Regan

Chapter 16
Author Reflections on the Human Factors of Transport Signs 239
Tim Horberry, Cándida Castro, and Patricie Mertova

Author Biographies 247

Index 255

1 An Introduction to Transport Signs and an Overview of This Book

Tim Horberry, Cándida Castro, Francisco Martos, and Patricie Mertova

CONTENTS

1.1 Introduction ... 1
1.1.1 What Is a Transport Sign? ... 2
1.2 International Traffic Signs ... 2
1.3 Requirements of Transport Signs ... 4
1.4 Transport Signing Research Issues: A Taster ... 6
1.4.1 Warning vs. Indication Signs ... 7
1.4.1.1 Warning Signs ... 7
1.4.1.2 Hierarchy of Hazard Control ... 7
1.4.1.3 Indication Signs ... 8
1.4.2 Text-Based vs. Symbolic/Pictorial Signs ... 9
1.4.3 Conclusions ... 12
1.5 This Book: Why Studying Transport Signs Is Worthwhile ... 12
1.5.1 The Structure of This Book ... 13
References ... 14

1.1 INTRODUCTION

Transport signs are ubiquitous; a quick look along almost any highway, rail line, or airport ramp will confirm this. The directness with which they communicate, and their exact designs, can almost be thought of as a form of art. As the next chapter in this book shows, these signs have a long history, easily predating motor cars, railways, or aviation. However, little serious research to improve their effectiveness had been undertaken before about 1960. Since the 1960s, admittedly, a large amount of research has been performed on transport signs; however, much of this work has been restricted to piecemeal testing of individual factors of signs (e.g., the most effective font type or the most effective sign to warn of a particular hazard). Overall, little work has been done that brings together much of the important work on signing

0-415-31086-5/04/$0.00+$1.50

from the perspective of the human operator, whether driver, pedestrian, or pilot. This book attempts to redress this imbalance.

1.1.1 What Is a Transport Sign?

Although seemingly obvious, for the purposes of this book it is necessary to define exactly what is meant by the term *transport/traffic sign*. According to the International Commission on Illumination (CIE),

> A sign is a device that provides a visual message by virtue of its situation, shape, color or pattern and sometimes by the use of symbols or alpha–numeric characters ... This short message is used to transfer information. The objective of such a communication is to have the receiver understand what the sender means. The sender is the sign designer and he must consider the discriminative, interpretative and recall skills of the driver and the environmental condition.
>
> **CIE, 1988, p. 3**

Slightly more recently, traffic signs have been defined as "an integral part of the road environment that can include not only upright signs giving warnings and instructions to traffic, speed limits, directions and other information, but also road markings, traffic light signals, motorway matrix signals, zebra and pelican crossings and cones and cylinders used at road works" (U.K. Department of Transport, 1991, p. 4). It is now also necessary to include in the definition traffic signs with a variable message. Furthermore, as will be seen later in this book, transport signing is not just restricted to roads; other transport modalities such as aviation and rail also have significant signing issues (See Figure 1.1).

1.2 INTERNATIONAL TRAFFIC SIGNS

With greater international travel by road, rail, or air, the desire is understandably to achieve greater consistency in the rules governing traffic movement. Looking purely at road transport in Europe, meetings such as the Convention on Road Signs and Signals held in Vienna on November 8, 1968, and subsequent amendments to the convention (e.g., on November 30, 1995) have tried to attain such uniformity (United Nations Economic Commission for Europe, 2003). The contracting parties in this Convention recognize that international uniformity of road signs, signals, symbols, and road markings is necessary to facilitate international road traffic and to increase road safety.

The system prescribed at this Convention differentiates between the following classes of road traffic signs (United Nations Economic Commission for Europe, 2003):

- *Danger warning signs* warn road users of a danger on the road and notify them of its nature.
- *Regulatory signs* notify road users of special obligations, restrictions, or prohibitions with which they must comply. These are subdivided into

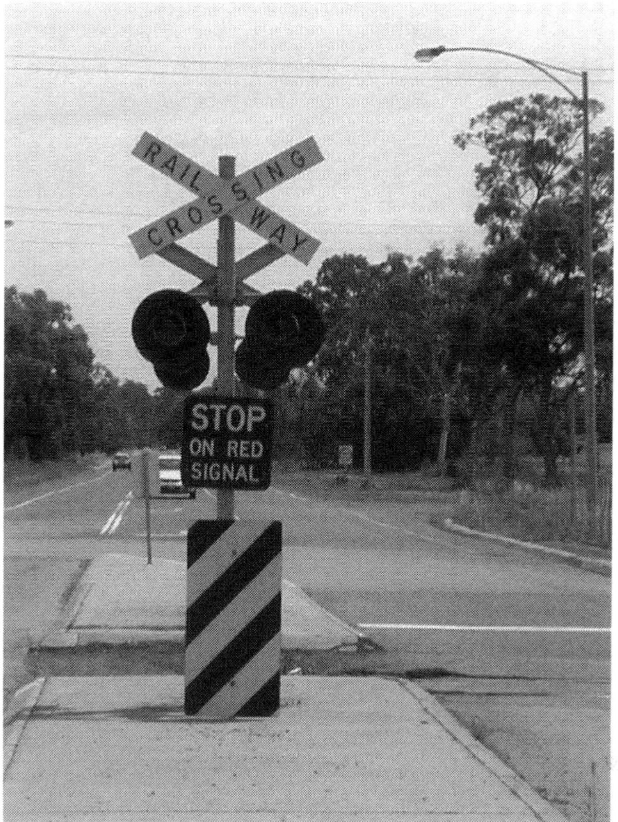

FIGURE 1.1 Signing that warns of the intersection of a road and railway.

priority signs, prohibitory or restrictive signs, mandatory signs, and special regulation signs.

- *Informative signs* guide road users while they are traveling or provide them with other useful information. These signs are subdivided into information, facility, or service signs; direction, position, or indication signs (including advance direction signs and road/place identification signs); and other panels.

In addition, the *Manual on Uniform Traffic Control Devices* (MUTCD) is an important document regarding international road signs. The latest version is the 2003 Millennium Edition (http://mutcd.fhwa.dot.gov). Published by the U.S. Federal Highway Administration, this manual is for use by traffic authorities, jurisdictions, and persons involved with the manufacture, installation, and maintenance of traffic signs on streets and highways in the U.S. To a large degree, its influence extends worldwide.

Despite this, very diverse designs of road traffic signs (in terms of their shapes, color, language, bordering, and so on) are used in different countries around the world (for instance, Europe, Japan, North America, and Australia). Idiomatic reasons, local criteria, resistance to change, and political issues help prevent the desirable outcome of full standardization of traffic signs. Progress has been made on some fronts; for example, the European Union countries are reaching some measure of agreement. However, as will be seen later in this book, new technologies such as Variable Message Signs and other Intelligent Transport Systems face similar standardization issues.

Many examples of "unique" signs may be taken from real traffic environments around the world. For instance, the two warning signs from Australia and Spain shown in Figure 1.2a do not exist in other countries. Figure 1.2b shows similar signs, in slightly different designs in England and Spain — another example of this lack of standardization. In locations with many international drivers, this lack of standardization is a particular problem, especially in terms of sign comprehension (Al-Madani and Al-Jahani, 2002). This topic will be tackled later in this book.

1.3 REQUIREMENTS OF TRANSPORT SIGNS

To be effective, signs in all transport modalities must at the very least be noticed, comprehended, and followed (or, in other words, seen, understood, and heeded; see Chapter 4 for a more detailed analysis). In some cases, poor acquisition and interpretation of the information conveyed by traffic signs can increase accidents (Horberry, 1998); however, better conformity to traffic signs can be achieved only partially by making signs more visible. A substantial number of industrial accidents and product liability incidents (reported in the safety science literature) show that the injured party observed and understood the warnings and safety instructions but chose to disregard them for reasons that seemed appropriate at the time (Zeitlin, 1994). Therefore, it is too simplistic to assume that if signs were more conspicuous they would automatically be better obeyed and the accident rate thus reduced.

Nevertheless, it is also true that when the visual aspects of a sign are particularly poor, improvements could possibly result in a quantum improvement in performance. Therefore, continuing effort must be made to increase the conspicuity and legibility of transport signs. As such, the elucidation of factors that determine the information that drivers select from their visual environments is an interesting subject of practical consequence.

Engels (1971) operationally defined visual conspicuity as the combination of properties of a visible object in its background that attracts attention via the visual system and is seen as a consequence. Cole and Jenkins (1982) and Cole and Hughes (1990) redefined conspicuity of the sign as the probability that the sign will be noticed by an observer within a fixed time or, conversely, as the time that an observer needs to notice the sign. Looking further at the issue of vision and attention, Hughes and Cole (1986) stated that drivers' visual attention is often attracted by advertisements and by other "irrelevant" objects in those sections of the route where advertising frequency is low. When analyzing the eye movement data, only about 15 to 20% of drivers' fixations are focused on the traffic devices. Similarly, Horberry (1998) found that for various warning and information signs, the percentage of drivers who actually looked at these signs was as low as 4% (and even for the best sign, only approximately 50%). Therefore, it is not certain that drivers notice all or even most traffic devices.

FIGURE 1.2a (See color insert following page 154.) Warning signs in Australia and Spain in the traffic environment.

Even when a sign is noticed, a driver still needs sufficient time to read it. Mori and Abdel–Halim (1981) estimated a range of 0.271 to 0.784 sec to read a road sign. This time depends on traffic conditions such as the distance of the sign and its position. This issue is examined further in Chapter 3.

Comprehensibility is another main characteristic that a sign must possess. Drivers must easily understand the message given by a sign, and the response required by the message must be clearly conveyed (Cooper, 1989). Cooper surveyed drivers' and nondrivers' comprehension of signs and found high comprehensibility levels for some signs currently used. For instance, the height limit sign was fully understood by 95% of U.K. drivers. On the other hand, other signs were not understood so well; examples included signs with more abstract symbols in them.

Anglo-Saxon and Spanish version
of obligatory direction sign
Anglo-Saxon and Spanish version
of double bend sign

FIGURE 1.2b (See color insert following page 154.) Examples of the lack of harmonization in traffic sign design around the world.

As well as being comprehensible, a sign must also be legible. The message displayed in the sign, whether made up of alphabetic characters or symbolic pictures, must be legible at the minimum distance from which it is to be read. This topic will be considered at more length in Chapter 3 and elsewhere in the book. Emerson and Linfield (1986) carried out a literature review on the legibility of traffic signs and highlighted several factors as determinants of traffic sign legibility: first were factors related to the design variables of the sign (such as character size, spacing between characters, character form, matrix format, light output, contrast, sign's conspicuity, and amount of information displayed); second were human factors (such as visual acuity and age); and last were environmental factors (e.g., night viewing, fog, and dusk).

Finally, a sign must be credible and accurate; the message conveyed in it should be credible (and convincing) to the reader so that he will act upon it (Wogalter et al., 1994; Edworthy and Adams, 1996). Credibility is related to drivers' motivation and is essential for eventual compliance with the sign's message. This is surely the crux of the issue of signs in relation to transport safety. What is it that makes a sign credible? What makes people comply with a sign? To what extent is it a human risk-taking problem rather than a visibility/perceptual design problem? Because many visibility/perceptual problems are indeed associated with signage design, it is easy for experts to dismiss the role of human risk-taking in sign compliance. The complexity of this conundrum will be tackled in several chapters later in this book.

An ideal transport sign should therefore fulfill at least these requirements: conspicuity, comprehensibility, credibility, legibility, and accuracy. Other issues, such as aesthetics, cost, and location of the sign, are generally also of critical importance. Throughout this book it will be argued that, usually, the most important measure to judge the effectiveness of a sign is its influence on the behavior of drivers for whom it has been provided (including its effect on accident rates).

1.4 TRANSPORT SIGNING RESEARCH ISSUES: A TASTER

This section browses through a sample of topics that help define traffic sign research. As will be seen throughout the book, a great deal of useful, established, and relevant data about signing is currently available, but these data still need a great deal of organizing and systematizing before they can be applied to design and installation of effective signs for all situations. This is especially the case with new signing technologies described toward the end of this book.

1.4.1 Warning vs. Indication Signs

To help organize the world of signing, a general taxonomy will be used here. There are an amazingly large number of different types of transport signs; to make sense of them, it is necessary to divide them by their different attributes. One method is to classify traffic signs by the messages they give, according to their different semantic categories (as was previously mentioned when discussing international traffic signs). For example, in the road environment, the U.K. Highway Code (e.g., Department of Transport, 1996) has three main categories: signs giving orders (e.g., "Stop"), warning signs (e.g., "Junction"), and direction and information signs (e.g., "Glasgow 60 miles"). They have three different basic shapes: triangular for warning signs, circular for those giving definite instructions, and rectangular for information signs. Much of the published work on traffic signs has focused on warning or indication/information signs. The next section will highlight some of the most important topics tackled and some of the most relevant conclusions obtained.

1.4.1.1 Warning Signs

Wogalter and Laughery (1996) state that the purpose of warning signs is twofold: to inform of potential hazards and to change behavior, that is, to stop unsafe acts. In fact, in an earlier report, Wogalter (1994) even suggested that the behavioral function was more important than the informational, in that it is more important to avoid the hazard than to know about it and then still be involved in an accident. Before this, Lehto and Miller (1986, p. 89) described the ultimate function of a warning. They asserted, "Even if a warning is perceived and comprehended, it will not be effective unless it induces people to behave safely."

Direct behavioral testing of a warning sign is, however, problematic. Wogalter and Laughery (1996) note some of the difficulties of performing such research. First, direct behavioral observation of warning effects can be time and labor intensive due to the infrequency of critical events (e.g., unsafe behavior). Second, allowing hazardous situations to occur must raise ethical concerns. Finally, laboratory studies, which allow good control, may not be applicable to other settings because of the difficulties of creating credible risk situations (i.e., the "ecological validity" may be low).

It appears that traffic warning signs generally only partially influence safe driving behavior, especially if the driver is familiar with the situation, is not sufficiently motivated to comply, or detects the situation too late, in which case the effect of the warning will be diminished. Indeed, Drory and Shinar (1982) found that fewer than 10% of drivers registered general traffic warning signs under a variety of roadway conditions. From this they concluded that "... under normal daylight conditions warning signs are either redundant (contain information directly available) or irrelevant to the driver's perceived needs and the driving task" (Drory and Shinar, 1982, p. 25).

1.4.1.2 Hierarchy of Hazard Control

Looking further at transport signs that warn of a potentially unsafe situation ahead (e.g., wild animals on the road/rail track), it must be pointed out that in many ways such signs are the weakest form of hazard control. How can a potentially hazardous situation (such as wild animals on the road/rail track) be stopped or reduced? A commonly held view in human factors and safety management (e.g., Horberry, 1998) is that there is a hierarchy of ways of doing so, for example:

- *Remove the problem* by designing it out — for example, build roads/rail lines away from locations where wild animals may be (obviously not always a possible option).
- *Place a barrier around the object* to stop the problem from occurring — for example, place a barrier/guarding around the roads/rail lines to stop wild animals from accessing them.
- *Warn of the danger* to try to induce safer behavior on the part of the driver — for example, erect signs that warn drivers of wild animals possibly ahead.

This is termed a hierarchy because, where possible, efforts should be made to address the hazard problem as high up the hierarchy as possible. However, in terms of expense, often the hierarchy is implemented in reverse: placing a few warning signs is often far cheaper than removing the problem by redesign (at least, for an existing roadway/rail line for which costly redesigns are proposed).

Of course, this does not reduce the importance of warning signs (and does not have an impact upon other types of signs, such as information signs). However, it offers a bit of perspective and shows that some types of signs, by their very nature, are not always the ideal solution. Perhaps, the low effectiveness rates for some warning signs found by some researchers are not that surprising because they are often not the optimal way of controlling potentially hazardous situations.

1.4.1.3 Indication Signs

When on an unfamiliar road, perhaps the best route-finding method would be to take someone along who knows the way and can provide proper information to help make the right decisions at the right time. An ideal system of indication signs must fulfill the functions of giving the correct information and displaying it at the appropriate time (Moore and Christie, 1963). It is vital to locate traffic signs correctly in order to guarantee their visibility at the right distance, allowing drivers to make the right maneuvers comfortably, day or night. Of course, this also applies to some specific warning signs, such as "Sharp Bend Ahead," and less to general warning signs, such as warning of animals on the road/rail line (See Figure 1.3).

Agg (1994) highlighted information overload as the main trouble that direction signs face in dense traffic systems. Direction sign information overload can happen when a sign has more destinations than can be read in the time available. Acquisition of information takes a certain amount of time and drivers should be able to read the traffic sign as quickly as possible in order to carry on in their main driving tasks (see Chapter 3).

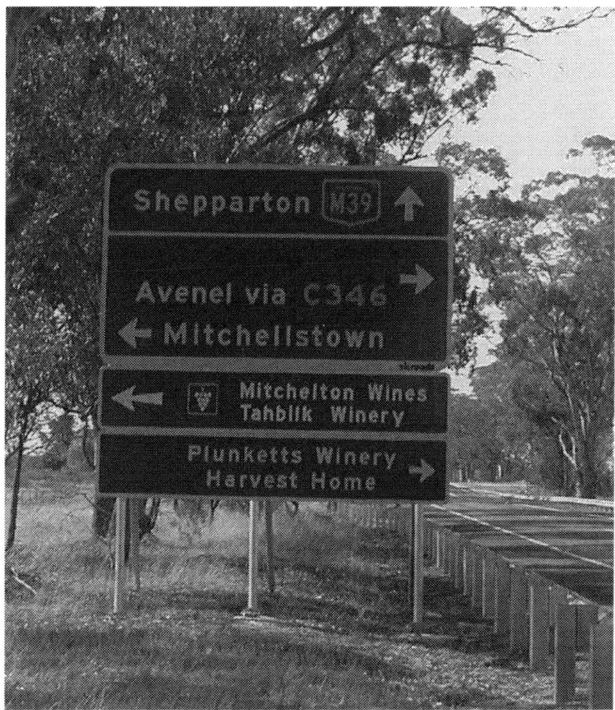

FIGURE 1.3 An indication sign giving directions.

1.4.2 Text-Based vs. Symbolic/Pictorial Signs

Over the past 40 years, a large amount of traffic sign research designed to look for the best possible forms of giving information has been carried out, in particular that concerning whether a symbolic version of a sign is more effective than a text-based one. This large interest is partly due to the increased use of symbolic signs in Europe and to some extent in the U.S. (Edworthy and Adams, 1996).

Early studies usually showed that icons were better than text. Jacobs et al. (1975) explored the effect of presenting alphabetic or symbolic traffic signs. Their results showed that symbolic signs could be legible from a threshold distance double that of alphabetic signs for all levels of visual acuity. Laboratory research by Ells and Dewar (1979) found that general traffic warning signs (e.g., “Hill Ahead”) with symbolic messages were understood more quickly by drivers than those solely with text-based messages. Furthermore, they found that when the signs were visually degraded, performance decrease was greater for text-based than for symbolic signs. This was later supported by Kline et al. (1990), who, when using general warning signs (e.g., “Road Narrows”), found that visibility distance for the icon-type (i.e., symbolic) traffic signs was larger than for text-type signs for young, middle-aged, and older subjects.

MacDonald and Hoffmann (1991) also showed the superiority of symbolic signs over text in terms of their conspicuity, legibility, and ease of comprehension. For instance, they found that drivers reported a significantly higher level of sign information in a driving field experiment with various symbolic signs along their route as compared with text-based signs. In the related area of product warning signs, Laughery and Young (1991) obtained similar results regarding the success of adding a pictogram or an icon to a sign. According to Edworthy and Adams (1996), the main advantages of symbolic over text-based signs are that symbolic signs:

- Can be recognized by those who do not or cannot read the language.
- Have a greater recognition distance.
- Are often recognized more quickly and more accurately than words.
- Can withstand greater degradation and still be recognizable.
- May be more effective when used with text in the same sign than text alone.

In addition, other data in favor of traffic sign symbolic presentation come from gerontology studies. Increasing age produces a significant reduction in visual acuity that leads to profound difficulties reading highway sign messages (Evans and Ginsburg, 1985). This handicap is further highlighted when light is reduced, such as driving at night or in bad weather conditions (Kline et al., 1992). This issue will be explored further in later chapters of this book. Older drivers require more time to process information and make decisions (Lerner, 1994) and are overrepresented in accident statistics. That is, they are involved in a comparatively higher percentage of traffic accidents than their middle-aged counterparts (Federal Highway Administration, 1989). Therefore, finding alternative ways to facilitate traffic sign information acquisition will become imperative as societies become progressively more elderly.

Symbols/pictograms are becoming increasingly important in our environment as an immediate source of information, although it must be noted that their usage is sometimes problematic. Kline et al. (1990) found considerable variation in comprehensibility from one icon to another. An additional problem is to decide the precise design of the symbol to be used in the sign (Cole and Jacobs, 1981), especially in complex situations. Using an example from one author's (Horberry, 1998) Ph.D. thesis, should a low bridge warning sign display a low bridge, a high-sided vehicle, a bridge and a vehicle, or an abstract design? As pointed out by Edworthy and Adams (1996), some complex situations cannot be captured in symbolic form without a good deal of appropriate learning. For instance, they give the example of the complex situation "Look up! Low door!" that cannot easily be displayed in symbols. Edworthy and Adams (1996) argue that the way around such problems is to have any proposed symbol appropriately tested; they recommend comprehension testing using appropriate subject groups (i.e., drivers) as a major part of this evaluation procedure.

Furthermore, in terms of driver preferences, symbolic signs are not always popular. For instance, Robertson (1977) compared unfamiliar symbol signs with text-based ones. Drivers preferred six text-based signs warning of following too closely to other drivers to six new versions with complex symbolic signs. These differences may have been due to subjects' unfamiliarity with the symbols, and further testing after the symbolic signs had been employed for several months on the roads may have produced different results.

Finally, several studies have found that symbols are effective when combined with text presentation. Moore and Christie (1963) emphasized the benefits of using words in combination with abstract traffic signs. The addition of words to symbolic traffic signs (without reducing their visibility) can be helpful to drivers who cannot fully understand their symbolism. Young (1991) pointed out that adding a pictogram or an icon to a warning sign reduced the time required to find and recognize the warning, thus enhancing the "noticeability" of the warning information. However, as a word of caution, some studies have found that adding a second code to a traffic sign (for example, combining a symbol and words) could make the message interpretation more difficult, especially in terms of speed of response or comprehension level (Avant et al., 1996; Horberry, 1998). Thus, it is difficult to come to an overall and firm conclusion regarding combining text with a symbol in a sign (See Figure 1.4). In addition, a similar debate about text/symbols for Variable Message Signs (VMS) is currently taking place. This is considered in more detail in later chapters.

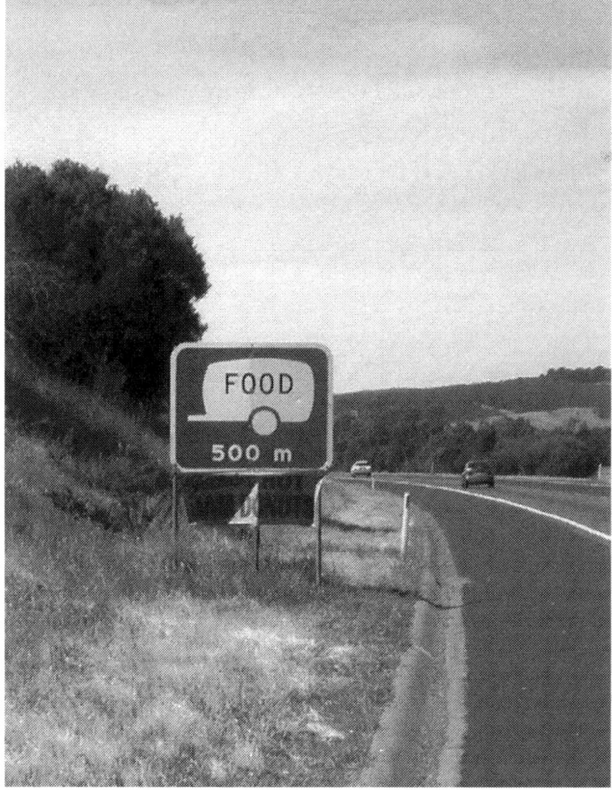

FIGURE 1.4 Example of a sign containing symbol and text.

1.4.3 Conclusions

The division of signs into warnings and indications is partly arbitrary. Nevertheless, one result common to both types is that they are usually more effective at night as compared with the day, when other sources of visual information are unavailable. The other frequently used distinction discussed here was that of symbolic vs. text-based signs; the weight of evidence is generally in favor of symbolic signs. However, the world is not always as simple; often there are situations in which text-based signs are essential (such as location names on indication signs). Thus, the authors are generally in favor of increased usage of symbolic traffic signs but slightly wary of attempts to symbolize almost everything. Well-designed textual signs can be useful in some situations, such as when precise numeric or instructional data are required or when a simple symbol cannot be identified. The seemingly perfect compromise of symbolic and textual information in the same sign is not always the panacea because mixed results of using this approach have occasionally been found.

Whether a sign is textual/symbolic or a warning/indication, it is the contention of the authors that the most important issue is for signs to be evaluated with an appropriate user group in realistic environments. Furthermore, it might be necessary to redesign the sign after the initial evaluation, and several iterations of the design-evaluation process should be performed before the sign is considered acceptable. The authors are not so naïve as to believe that a "perfect" sign is always possible, but they argue that appropriate testing and development to help create appropriate signs are crucial to the safety and efficiency of the transport system.

1.5 THIS BOOK: WHY STUDYING TRANSPORT SIGNS IS WORTHWHILE

> When man first began to move around his Earth, he was guided by nature; paths and trails often followed the contours of the land. Warning signs were provided by animal tracks or rushing water, by smells and sounds. There was no need of regulation by man. Imperial Rome provided road signs for travelers. Under Caesar Augustus, the 29 major military highways which led from the city to the outposts of the empire were provided with milestones for their first 100 miles. A law establishing compulsory measurement of these routes was enacted in 183 B.C. It took almost 200 years for a standard milestone to come in general use. Neither travel nor road signs changed significantly during the next 18 centuries.
>
> **Lees and Farman, 1970, p. 7**

Transport signs are undergoing a dramatic change. As the preceding excerpt shows, whereas Roman signs stayed in use for centuries and the standards established under Caesar Augustus are still to some extent valid, new technologies are currently revolutionizing signaling. Future signs displaying varying messages or Intelligent Transport Systems will be much more powerful and dynamic means of conveying information. Such power, however, does not lack risks. Signs aim at organizing some of the chaos of visual information that impinges on drivers' senses; however, signs themselves are another kind of stimulus that can increase the confusion. As an extreme example, a Variable Message Sign in the road environment that attracts the driver's attention at the entrance of a roundabout simply to remind him to renew his driving license might not be the best placement possible.

With the ever-increasing use of these new Variable Message Signs and Intelligent Transport Systems, one might wonder if it is, after all, worth studying transport signs. Of course, these authors argue that it is. The design of new signs and the choice of the messages they display must take into account existing knowledge about current signs in order to prevent them from jeopardizing, instead of improving, transport safety. In addition, traffic signs can be very cost effective; money spent on accident prevention through signing increases safety and efficiency of the road, rail, and aviation environments and can save lives, money, and effort (a topic considered further in Chapter 3 and Chapter 4).

Signs in some form will be located in transport environments for many years, perhaps even centuries. Because no general, easily accessible information source summarizes the main findings about transport signs from a driver-centered perspective, this book intends to offer such a guide. Even though it concentrates largely on the road environment and general signing issues, it will also address other transport modalities, such as the railways. One of our purposes is to make the relevant information easy to understand and use. Each chapter of this book discusses a specific issue regarding the interaction of transport signs and driver behavior and, where appropriate, includes a final synopsis regarding their practical implications.

1.5.1 The Structure of This Book

The overall structure of this book is intended to be as clear as possible, ranging from a description of general issues with regard to traffic signs to consideration of more specific topics. In Chapter 2, transport signs will be put into context by considering their history. Thereafter, the three chapters that follow will examine signing fundamentals: design issues (Chapter 3), the effectiveness of transport signs at a localized/behavioral level (Chapter 4), and the effectiveness of signing within the wider transport system (Chapter 5).

Because this book also considers transport signing in environments other than highways, the next two chapters will consider signing issues in railway operations, including the issue of signals passed at danger (Chapter 6) and aviation signing (Chapter 7). Following this, more specific signing issues will be considered: visual factors, aging, and signing (Chapter 8) and attentional and motivational factors involved in transport signing (Chapter 9). Subsequently, the focus slightly broadens to consider signing within a wider societal context by considering such factors as cross-cultural differences with respect to sign design and comprehension, legislation, and international standardization (Chapter 10 and Chapter 11).

New technologies exert a massive impact upon all transportation systems, so the final chapters of the book examine this topic. Chapter 12 and Chapter 13 consider Variable Message Signs in the road environment (Chapter 12 from a "bottom-up" perspective and Chapter 13 with a more "top-down" approach). Following this, Chapter 14 introduces Intelligent Transport Systems and considers how such technologies might replace (or at least change the role of) traditional transport signs. Chapter 15 is more speculative and considers a range of fundamental Human Factors implications with regard to these new technologies within transport systems as a whole. Finally, in Chapter 16 the authors and editors bring together chapter authors' reflections regarding their contributions to the book, along with their short biographies.

REFERENCES

Agg, H. (1994). Direction sign overload. Research report, 77, No, 15. Crowthorne: Transport Research Laboratory.

Al-Madani, H. and Al-Jahani, A.R. (2002). Role of drivers' personal characteristics in understanding traffic sign symbols. *Accident Anal. Prev.*, 34(2), 185–196.

Avant, L.L., Thieman, A.A., Zang, A.L., and Hsu, S.Y. (1996). Memory codes for traffic sign information: visual vs. meaning codes. In *Vision in Vehicles, V.* A.G. Gale, I.D. Brown, C.M. Haslegrave, S. Moorhead, and S.P. Taylor, Eds. Amsterdam: Elsevier.

CIE: International Commission on Illumination. (1988). Roadsigns. Vienna: Central Bureau of the CIE. Publication No. 74.

Cole, B.L. and Jacobs, R.J. (1981). A comparison of alternative symbolic warning signs for railway level crossings. *Aust. Road Res.*, 11(4), 37–45.

Cole, B.L. and Jenkins, S.E. (1982). Conspicuity of traffic control devices. *Aust. Road Res.*, 12, 221–238.

Cole, B.L. and Hughes, P.K. (1990). Drivers don't search: they just notice. In *Visual Research.* D. Brogan, Ed. London: Taylor & Francis.

Cooper, B.R. (1989). Comprehension of traffic signs by drivers and non-drivers. Research report, 167. Crowthorne: Transport and Road Research Laboratory.

DoT: Department of Transport (1991). Specifications for drawings and markings: warning signs. (P)530, DoT: London.

DoT: Department of Transport (1996). The highway code. HMSO: London.

Drory, A. and Shinar, D. (1982). The effect of roadway environment and fatigue on sign perception. *J. Saf. Res.*, 21, 25–32.

Edworthy, J. and Adams, A. (1996). *Warning Design: a Research Prospective.* London: Taylor & Francis.

Ells, J.G. and Dewar, R.E. (1979). Rapid comprehension of verbal and symbolic traffic sign messages. *Hum. Factors*, 21(2), 161–168.

Emerson, P.G. and Lindfield, P.B. (1986). Perception of variable message traffic signs. In *Vision in Vehicles, I.* A.G. Gale, M.H. Freeman, C.M. Haslegrave, P. Smith, and S.P. Taylor, Eds. Amsterdam: Elsevier.

Engels, F.L. (1971). Visual conspicuity, directed attention and retinal locus. *Vision Res.*, 11, 563–576.

Evans, D.W. and Ginsburg, A.P. (1985). Contrast sensitivity predicts age-related differences in highway-signs discriminability. *Hum. Factors*, 27, 637–642.

Federal Highway Administration. (1989). The federal highway administration's action plan for older persons. Washington, D.C.: U.S. Department of Transportation.

Federal Highway Administration (2003). Manual on uniform traffic control devices. http://mutcd.fhwa.dot.gov. Retrieved January 21, 2004.

Horberry, T.J. (1998). Bridge strike reduction: the design and evaluation of visual warnings. Unpublished Ph.D. thesis, University of Derby, U.K.

Hughes, P.K. and Cole, B.L. (1986). What attracts attention when driving? *Ergonomics*, 29(3), 377–391.

Jacobs, R.J., Johnston, A.W., and Cole, B.L. (1975). The visibility of alphabetic and symbolic traffic signs. *Aust. Road Res.*, 5(7), 68–73.

Kline, T.J.B., Ghali, L.M., Kline, D.W., and Brown, S. (1990). Visibility distance of highway signs among young, middle-aged and older observers: icons are better than text. *Hum. Factors*, 35(5), 609–619.

Kline, D.W., Babbitt. T.J., Fozard, J., Kosnik, W., Schieber, F., and Sekuler, R. (1992). Vision, aging and driving: the problems of older drivers. *J. Gerontol. Psychol. Sci.*, 47, 27–34.

Laughery, K.R. and Young, S. (1991). An eye scan analysis of accessing product warning information. *Proceedings of the Human Factors Society 35th Annual Meeting*. Human Factors Society: San Francisco.

Lees, J. and Farman, M. (1970). An investigation of the design of performance of traffic control devices. *Visible Language*, 4, 7–38.

Lehto, M.R. and Miller, J.M. (1986). *Warnings: Volume I: Fundamentals, Design and Education.* Ann Arbor, MI: Fuller Technical Publications.

Lerner, N. (1994). Giving the older driver enough perception–reaction time. *Exp. Aging Res.*, 20, 25–33.

MacDonald, W.A. and Hoffmann, E.R. (1991). Drivers' awareness of traffic sign information. *Ergonomics*, 34(5), 585–612.

Moore, R.L. and Christie, A.W. (1963). Research on traffic signs. Road research laboratory. *Proc. Eng. Traffic Conf.* July, 113–122.

Mori, M. and Abdel–Halim, H. (1981). Road sign recognition and nonrecognition. *Accident Anal. Prev.*, 13, 101–115.

Robertson, A. (1977). A road sign for warning of close following: form and message design. TRRL supplementary report 324. Transport and Road Research Laboratory: Crowthorne.

United Nations Economic Commission for Europe (2003). Convention on road signs and signals (2003). /http://www.unece.org/trans/conventn. Retrieved January 21, 2004.

Wogalter, M.S. (1994). Factors influencing the effectiveness of warnings. In *Proceedings of Public Graphics*. H. Zwaga, Ed. University of Utrecht: The Netherlands.

Wogalter, M.S., Jarrard, S.W., and Simpson, S.N. (1994). Influence of warning label signal words on perceived hazard level. *Hum. Factors*, 36(3), 547–556.

Wogalter, M.S. and Laughery, K.R. (1996). Warning! Sign and label effectiveness. *Curr. Directions Psychol. Sci.*, 5(2), 33–37.

Young, S.L. (1991). Increasing the noticeability of warnings: effects of pictorial, color, signal icon and border. *Proceedings of the Human Factors Society 35th Annual Meeting*, 580–584. Human Factors Society: San Francisco.

Zeitlin, L.R. (1994). Failure to follow safety instructions: faulty communication on risky decisions? *Hum. Factors*, 36, 172–181.

2 History of Traffic Signs*

Maxwell G. Lay

CONTENTS

2.1 Route Signing 17
2.2 Warning Signs 18
2.3 The Stop Sign 19
2.4 International Signs 20
2.5 Materials 21
2.6 Maritime and Rail Influences 21
References 22

2.1 ROUTE SIGNING

Routes along early paths were often marked with broken twigs, sticks, or stones. More elaborate waymarkers were signs carved on rocks, pieces of vertical stone called *stoops*, stones piled into cairns, and marked trees. One variant encountered on prehistoric English ridgeways was to line the way with the burial mounds or barrows of local chieftains. The practice was also common among the Greeks and Romans, who often buried their dead in graves beside the approach roads to their towns and cities.

For distance markers, or mileposts, the Romans used 600- × 600-mm marble shafts up to 3 m high and at 1.5-km spacing. The practice began with the Tribune Gaius in 123 B.C. and about 200 years after the Romans had commenced building arterial roads. The geographic origin of the network was a centerpiece in Rome in the form of a pillar erected by the Emperor Augustus in 20 B.C. at a corner of the Forum. Its bronze plaques displayed the distances to the empire's key towns. Later, it was referred to as the golden milestone, with gold lettering. Within about 200 km of this counterpoint, the roadside posts showed the distance to Rome; elsewhere, they referred to the distance to the nearest large town. They all also carried political messages. Often milestones were unnecessary. For example, smoke was always a daytime sign of a nearby town. At night, a number of towns lit lights or blew horns to guide travellers to their facilities.

As traffic increased, signing grew in importance. Mathew Simons' 1635 *Directions for English Travillers* noted that directional signs were found "in many parts

---

* The material in this chapter is largely based on Chapter 6 of the author's *Ways of the World* (see 1992 and 1993 listings in reference list), where more detailed references will be found.

0-415-31086-5/04/$0.00+$1.50

where wayes be doubtful." Early in the 17th century, the Duc de Sully and Cardinal Richelieu introduced a system of destination signing at French crossroads using cairns or posts. In 1669 Louis XIV formalized the system in a royal order. In 1697 Ole Römer in Denmark instituted an extensive network of mileposts at 1.8-km spacings.* A year later a new English law required each parish to place guideposts at its crossroads. Finger posts, often shaped like a pointing hand, and stone mileposts set in the ground became commonplace from the 18th century. Indeed, the development of toll roads and the postal service in that century led indirectly to great improvements in signposting, and some signs even included travel times. By 1835 English law quite specifically required universal direction signing.

In the U.S., systematic route marking began with a 1704 Maryland law requiring trees beside a route to be marked with an elaborate system of notches, letters, and/or colors. The system could indicate whether a road led to a ferry, church, or courthouse. One South Maryland road between the Potomac and Patuxent rivers has retained the name Three Notch Road. A different Three Notch'd Road ran from Richmond, Virginia, to the Shenandoah Valley from 1730 to 1930, after which it became part of U.S. Route 250. In the 18th century, signposts were often funded by public subscription, and in the 1760s Ben Franklin actively promoted milestones as an aid to his postal service (Sessions, 1976). Road signage subsequently came under police protection in the 19th century. In the early days of the bicycle and the car, the Maryland influence prevailed and major American routes were often distinguished by coded bands of color painted on roadside poles and trees.

At slow travel speeds it was always possible to ask someone for directions. At the end of the 19th century, the car was to render this procedure impractical. In 1894 the French Touring Club (Touring-Club de France) began erecting signs to aid its members. Between 1895 and 1896 the Italian Touring Club erected some 140 cast-iron signs carrying directional arrows, mainly on the coast road near Senigallia. In 1901 the Automobile Club of America began signposting roads with cast-iron strip signs. During the 18th century, the French began numbering their national routes for administrative purposes, and in 1913 André Michelin lobbied successfully for public route numbering. In 1914 the Swiss formed a society for route numbering and by 1915 had erected several thousand signs (Krampen, 1983). Route numbering began in the U.S. in Wisconsin in 1918. National route numbering was introduced by the American Association of State Highway Officials (AASHO) in 1925 after intense political and commercial debate and in association with the definition of a system of interstate highways (Sessions, 1976).

2.2 WARNING SIGNS

Modern road signs began in the late 19th century, when bicycle clubs erected many road signs. Two of the more common were "To Cyclists: This Hill Is Dangerous" and "Caution for Cyclists." As bicycle clubs developed into auto clubs, they often took their signing charter with them. Symbolic signs were introduced in Austria in 1910 to warn of locations where a steep slope demanded cautious braking. The sign

* Romer was also a successful astronomer who made the first realistic measurement of the speed of light.

consisted of drawings on roadside rocks of wagon wheel brake shoes. The incentive for the signs was a then-famous road death at Brennbuhel near Imst in Austria in 1854. King Frederick Augustus II of Saxony was killed when his coachman failed to use his brakes in just such a situation (Krampen, 1983).

The 6th Permanent International Association of Road Congresses (PIARC) World Congress (held in Washington, D.C., in 1930) strongly recommended warning sign use, eliminating words such as "turn" and "curve." This advice was heeded when the U.S. introduced its *Manual of Uniform Traffic Control Devices.*

2.3 THE STOP SIGN

In 1868 the British began using movable semaphores to signal "stop" on a road outside the Houses of Parliament in London. They were based on railway practice and, in the context of this book, are more similar to traffic signals than traffic signs (see further discussion in Section 2.6).

The octagonal stop sign arose out of this process, having been introduced in a simpler form in Detroit in 1914, when policeman Harry Jackson cut the corners off a square sign (McShane, 1999). It was used to halt cross-traffic on a boulevard leading into Detroit, and its installation was financed by the Motor Club of Michigan because the Detroit City Council refused to pay for it. The octagonal shape of the stop sign was formalized in 1923 at a meeting of the Mississippi Valley Association of State Highway Departments. Because the octagon shape required a lot of cutting and wastage, it was assigned to the stop sign, which was not expected to be used frequently. The sign was originally black on white and was changed to the current white on red in 1924 at the first U.S. National Conference on Street and Highway Safety. A period of "to-ing" and "fro-ing" between white on red and black on yellow followed. The 1927 AASHO rural sign manual specified "black on yellow." The first 1930 *Manual of Uniform Traffic Control Devices* (MUTCD) defined the octagonal sign as "red on yellow," carrying the words:

and the 1935 manual specified "black or red on yellow" with only the word "STOP" (Hawkins, 1992).

In 1931 the League of Nations' Convention adopted the octagon stop sign, and it was first used in Europe by France in the 1930s. In Britain in the late 1930s, traffic authorities still used a red triangle inside a red circle with the legend "Halt at Major Road Ahead" underneath (How, 1936). The current white on red was adopted in the U.S. in 1954, after improved red paints came on the market.

2.4 INTERNATIONAL SIGNS

The first attempt at international traffic signs occurred in the late 1890s at a London meeting of the international League of Tourist Associations, which discussed a proposal based on Italian arrow signs. Although the matter remained topical among tourist bodies and automobile clubs, formal government action did not occur until 1908 at the initial meeting of the international road group, PIARC. The resulting PIARC signs were first used for a road race in Austria in 1909 and ratified at an international convention later that year (Krampen, 1983). The four standardized signs were for gutters across roads, sharp curves, crossroads, and railway crossings.

Pressures from the 1921 International Traffic Congress and the 1923 PIARC World Congress to amend the 1909 convention led to a League of Nations study beginning in 1923 and ending with a convention being adopted in 1926. The additional signs in this convention warned of an unprotected railway crossing and of a general need for caution. The last sign was a red triangle based on a sign introduced in England in 1904 but — as in 1909 — all the other signs could be of any color. The U.S. adopted the triangular yield sign in 1971, although a black on yellow sign with the words:

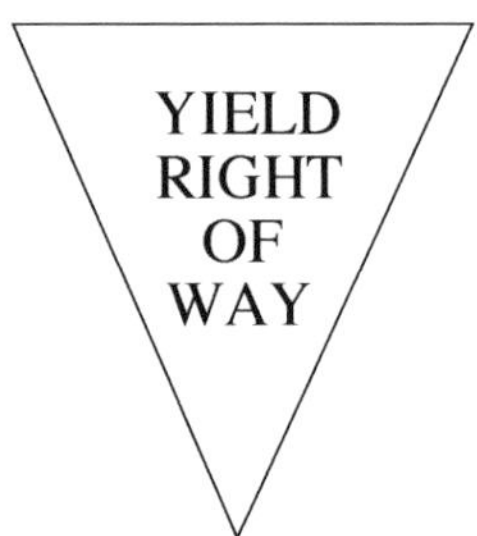

was introduced in the 1954 MUTCD. The message was shortened in the 1961 MUTCD and the 1971 edition changed the colors to today's red on white.

White on a blue background was the most commonly used color combination on signs, following a practice introduced in France in 1835 that used dark blue (often enamelled) cast-iron signs with white lettering (Wouvier, 1988, 1992). Nations progressively endorsed the convention between 1929 and 1939, with Britain being one of the last signatories. The British were initially separated from the League of Nations convention by a desire to accompany their signs with written messages. Standard signing in Britain began with the Local Government Board in 1904, but uniform signing was not introduced until 1921.

A League convention in Geneva in 1931 was devoted to further standardization of road signs and expanded the set to include the "! Attention" sign, the triangular right of way sign from French practice, and the abstract "No Entry" sign using a horizontal white bar on a red disc. This last sign drew on Swiss experience, which in turn was based on the practice adopted by some earlier European states of marking their boundaries with their formal shields and tying a blood-red ribbon horizontally around the shield when they did not wish visitors to enter.

The first American signing manuals were produced by Ohio and Minnesota in 1921. Signing practice was codified nationally by AASHO in 1924 and 1927, drawing on existing state systems. A *Manual of Uniform Traffic Control Devices* (MUTCD) was first produced in 1935, with a second edition in 1949 (Hawkins, 1992).

Early signing practice is reviewed in Sessions (1976). The large blue and white freeway-style direction signs with their distinctive lower-case letters naming each direction were introduced during the 1930s on the German autobahnen, but they had their antecedents in the earlier blue and white enamelled signs introduced by the French (Nouvier, 1988). In order to hinder the local progress of foreign armies, many European signs were destroyed during the two world wars. After the Second World War, the United Nations took over international sign standardization and continued the practice of producing signing conventions, beginning with a conference in Geneva on road and motor transport in 1949. The outcome was the Protocol on Traffic Signs and Signals, which retained all the old warning signs except the caution triangle and added 14 more.

By the time this protocol was issued, three different signing systems were effectively in use around the world: the European system, which was close to the Protocol; an African system that was a combination of the Protocol and earlier English practice; and a Pan-American system founded on the U.S. manual and relying more on words than on symbols. The reliance on words reflected the fact that the U.S. system needed to accommodate only one official language. Although called Pan-American as a result of decisions at Pan-American Road Congresses in 1939 and 1949, its English basis meant that it had little impact in South America and was effectively the U.S. system.

In 1968 the U.N. Convention in Vienna proposed a truly international system based on an acknowledgment that the European and American systems were of equal value. Thus, the number of warning signs increased from 24 at Geneva to 41 at Vienna. Today, the world still lacks an international system of traffic signs and two major systems remain in use — the European and the American — with European symbols gradually becoming the world standard.

2.5 MATERIALS

The usefulness of signs was greatly enhanced by use of reflective sign surfaces. Reflective lenses were used on railway crossing gates in Switzerland from 1918 and reflectorized road signs were introduced in 1921 with the use of ribbed glass panels. The panels were supplanted by glass-bead technology following the development of the beads by Rudolph Potters in 1914 and Micheli's application of the technique in 1925. Luminous paints have been in use since World War II.

2.6 MARITIME AND RAIL INFLUENCES

At the time of the Spanish Armada in 1588, the news that the sailing ships of the Armada were in sight was conveyed across the English countryside by huge bonfires set on a string of strategic hilltops. When a similar threat was posed by the forces of Napoleon, the advances of the Industrial Revolution meant that the bonfires were replaced between Portsmouth and London by a series of mechanically operated semaphores placed on signal posts at the top of brick towers. These semaphores on posts were soon adopted for the new steam-powered railways in the 1830s. To be visible at night, the semaphore arms needed lights to replace the message conveyed in daytime by the semaphore orientation; red, yellow, and green were used, based on maritime precedents. The somewhat flawed process of adoption is well told in Wilson (1855).

Signaling by flags and colors was a well-established maritime practice. Red was widely adopted for danger, probably because of its emotive association with blood and despite its having poor visibility (e.g., about $^1/_{10}$ that of white) due to the loss of luminance. That's why starboard (rightside) lights are green and port (leftside) lights are red on ships. With the coming of steam-powered ships, international maritime rules were first codified in 1863, a quarter of a century before the first practical car. Thus, the meaning of the red, yellow, and green signals now common in road signals comes directly from 19th-century maritime and rail practice. For example, common railway practice used a square lamp with red lenses in one direction and green lenses in the orthogonal direction. Rotating the lamp by 90° about a vertical axis would change the signal message from stop to go.

Traffic signals were discussed in Section 2.3 and, from the preceding discussion, it is not surprising to learn that the inventor of the first signals, which were placed outside the Houses of Parliament in London in 1868, was John Knight, who had been general manager of the London-to-Brighton railway line and had previously invented the electric lighting of train carriages. His traffic signals used red and green gas lamps and were based on the railway semaphores discussed earlier.

Red, in particular, became a key color; because many of the early railways were operated by staff from the Army and Navy, the system of signal, semaphores, and red flags became very regimented. The red flag entered road operations via Britain's infamous Red Flag Act that, from 1865 to 1896, required powered vehicles to be preceded by a man on foot carrying a red flag. For most of the 20th century, many pedestrian crossings were designated by red flags.

REFERENCES

Anon. (1936). *How to Drive a Car.* London: Temple Press.

Hawkins, H.G. (1992). Evolution of the MUTCD: early standards for traffic control devices and early editions of the MUTCD and the MUTCD since World War II. Institute of Traffic Engineers (ITE) J, July, pp. 23–26; August, pp. 17–23; November, pp. 17–23.

Krampen, M. (1983). Icons of the road. *Semiotica*, 281, 43$^1/_2$, 1–204.

Lay, M.G. (1992). *Ways of the World.* New Brunswick: Rutgers University Press.

Lay, M.G. (1993). The history of traffic signs. *PIARC Routes/Roads*, 281(III):83–88.

McShane, C. (1999). The origins and globalization of traffic control signals. *J. Urban Hist.*, 25(3), 379–404, March.

Nouvier, J. (1988 and 1992). Une histoire de la signalisation en France à les numéros de la RGRA & La signalisation avant 1920. *Revue générale des routes et des aerodromes* (RGRA), 658: 9–15, December; 699:17–20, September.

Sessions, G.M. (1976). *Traffic Devices: Historical Aspects Thereof.* Washington, D.C.: Institute of Traffic Engineers.

Simons, M. (1635). *Directions for English Travillers.*

Wilson, G. (1855). *Researches on Colour-Blindness; with a Supplement on the Danger Attending the Present System of Railway and Marine Coloured Signals.* Edinburgh: Sutherland and Knox.

3 Design of Traffic Signs*

Maxwell G. Lay

CONTENTS

3.1 Traffic Control Devices....26
- 3.1.1 Role of Traffic Control Devices....26
- 3.1.2 Traffic Sign Definitions....27
- 3.1.3 Traffic Sign Materials....27
  - 3.1.3.1 Retroreflective Sheet....27
  - 3.1.3.2 Variable Message Signs (VMS)....29

3.2 Traffic Sign Theory....29
- 3.2.1 General....29
- 3.2.2 Visibility....30
  - 3.2.2.1 Geometric Requirements....30
  - 3.2.2.2 Optical Requirements....31
- 3.2.3 Conspicuity....33
- 3.2.4 Detecting Detail....34
- 3.2.5 Detecting Whole Words....36
- 3.2.6 Glance Legibility....37
- 3.2.7 Understanding the Sign....38
- 3.2.8 Acting on the Sign....40
  - 3.2.8.1 Cognitive Response....40
  - 3.2.8.2 Physical Response....40

3.3 Traffic Sign Requirements....42
- 3.3.1 Shape and Color....42
- 3.3.2 Direction Signs....44
- 3.3.3 Safety....46

3.4 Conclusions....46

References....47

* The material in this chapter is largely based on Chapter 21 of the author's *Handbook of Road Technology*, Vol. 2 (see complete listing in References), where more detailed references will be found.

0-415-31086-5/04/$0.00+$1.50

3.1 TRAFFIC CONTROL DEVICES

3.1.1 Role of Traffic Control Devices

Traffic control devices include signs, signals, pavement markings, curbing, traffic islands, medians, and other installations provided for road users for the following purposes:

- *To instruct or direct road users*, i.e., to provide instructions or regulations that are required by law to be obeyed and might otherwise be overlooked. Such traffic control devices are called *regulatory devices* and failure to comply with them is an offense. An example would be a sign defining priority at an intersection. The two types of regulatory signs are:
  - *Prohibitory* (or *negative*) *signs* indicate a forbidden action (e.g., "No Entry") or restriction (e.g., "5 t Load Limit"). They are usually white discs with a red annular border. Prohibitory signs lead to slower response times and higher error rates than do mandatory (positive) signs (MacDonald and Hoffmann, 1978).
  - *Mandatory* and *permissive* (or *positive*) *signs* indicate an essential action (e.g., "Turn Left"), an instruction to proceed (e.g., "Through Traffic"), or an exclusive action (e.g., "Bikes Only"). They are usually colored discs with white symbols. Mandatory signs lead to shorter response times and lower error rates than do prohibitory signs (MacDonald and Hoffmann, 1978).
- *To identify relevant features*, i.e., to mark or warn of hazards ahead that would not otherwise be self-evident or expected; these are called *warning devices*. Typical examples are signs that warn of sharp curves, poor surfaces, intersections, advisory speeds, reduced clearances, and temporary roadworks. Warning signs are often characterized by white triangles with a red border. They can have a high benefit/cost ratio in terms of crash reduction.
- *To inform road users*, i.e., to convey useful information. They may be *permanent* information signs such as those giving navigational or locational data, informing and advising road users of such items as directions, distances, destinations, routes, points of interest, and location of services. These are called *guide* (or *information*) *signs*. They also include *temporary* information signs that give data on local traffic conditions, roadwork, road closures, diversions, and the like.

Traffic control devices are thus provided to aid in ensuring safe, predictable, efficient, and orderly movement of traffic. The message that they are intended to convey to drivers and pedestrians should be consistent with the other features of the road. Their specific use is usually governed by regulations and warrants developed by the relevant traffic authority. Four major risks with traffic signs are that they may:

- Cause a safety hazard (Section 3.3.3).
  - Cause visual blight (Blake, 1964).
  - Provide inappropriate information (Section 3.2.1).
  - Cause information overload (Section 3.2.6).

3.1.2 Traffic Sign Definitions

The purpose of this chapter is to deal with *traffic signs*, one form of traffic control device. The distinction between a traffic *sign* and a traffic *signal* (Lay, 1998, Chap. 23) has traditionally been that a sign passively conveys a static, unchanging message, whereas the message on a signal varies over time. Also, "passive" implies that communication is only between sign and road user and that no road user–sign interaction takes place. Now in many devices this distinction is blurred; this discussion will treat a traffic sign as the device that conveys an external visual signal to a driver.

It is essential that traffic signs be of uniform design, application, and location and that they convey uniform messages. This is needed to assist rather than confuse drivers, to reduce decision-making quality and time in unfamiliar situations, and, because traffic signs often perform a legal role, to withstand legal scrutiny. Decision-making time is critical because, for reasons that will become more obvious later in the chapter, a driver should be able to interpret the message on a sign correctly in one glance. There is no worldwide uniformity of traffic signing; the most widely followed document is an international Convention (or Protocol) on Road Signs and Signals produced by the United Nations in Vienna in 1968 (United Nations, 1952, Chap. 2). It is still maintained by the U.N. at a low level. It almost exclusively uses icons based on symbolic codes and a few textual messages and generally forms the basis for the work described in this chapter.

Traffic signs can be easily misused. For example:

- Signs may be contrary to driver expectations.
- Unusual signs can readily confuse drivers when they are making decisions under pressure.
- Signs may be expected to achieve more than is feasible.
- Poorly maintained signs may be difficult to interpret and may lead to drivers' disrespecting their messages.
- Signs may be obscured by environmental factors such as dust or smoke.
- Signs may be hidden by the road alignment.

3.1.3 Traffic Sign Materials

3.1.3.1 Retroreflective Sheet

Many signs convey fixed (or passive) "painted" messages. Today this is usually achieved by placing sheets of colored retroreflective material on a supporting background. This material has high specular (i.e., mirror-like) reflection properties usually achieved by inserting glass beads or other reflective microstructures into a transparent medium covering the sign surface. Retroreflective sheet is usually specified by the coefficient of luminance intensity (CIL), which is the ratio of the reflected luminance intensity to the illumination falling on the device per unit of sheet area (Lay, 1998, Chap. 24). This is sometimes called the specific intensity per area (SIA) and the units are cd/lx.m^2.

TABLE 3.1
Typical Values for New Sheet Viewed Almost Normal (0.2°) to the Surface and with a Light Source Close to Normal (4.0°)

	Specific Intensity per Area (SIA), or CIL per Area of Sheet, cd/lx.m²						
	Sheet Color						
Sheet Quality	Silver–White	White	Yellow	Red	Standard Green	Blue	Brown
High	220	250	170	45	24	20	12
Average	85	55	50	16	10	7	4

Retroreflective sheet is usually specified using the CIL approach developed for corner-cube reflectors; however, in this case it is normal to use the CIL per area of sheet, i.e., cd/lx.m² = cd/lm, rather than CIL per delineator. This is called the *specific intensity per area* (*SIA*). SIA values depend on the angle at which the observation is made. Typical values for new sheet viewed almost normal (0.2°) to the surface and with a close-to-normal (4.0°) light source can be seen in Table 3.1.

Altering the viewing angle from 0.2 to 0.33° cuts the CIL for white, for instance, from 250 to 180 cd/lx.m²; similarly, altering the light source angle from 4 to 30° cuts the CIL for white from 250 to 150 cd/lx.m². It is commonly assumed that the maximum angle at which a retroreflective sign must be seen is 30° from the direction of travel. Table 3.1 indicates the losses incurred using colors other than white or yellow (see also Section 3.3.1).

Section 3.2.2.2 will indicate that sign luminance levels should be between about 30 and 100 cd/m², with an optimum level of around 80 cd/m², and that a minimum luminance of 3 cd/m² is needed to ensure night-time legibility of signs; significant decrements occur below 30 cd/m². Based on the preceding data, retroreflective sheets can be seen to meet these needs. For example, the illumination provided by headlights is between 0.01 and 0.25 lx (Lay, 1998, Chap. 24), so a CIL of about 200 cd/lx.m² would result in a sign luminance of between 2 and 50 cd/m².

The sheets deteriorate with time and usually need replacement when they have lost 80% of their retroreflectivity. This typically happens after 8 to 15 years. In particular, a high-class white sheet loses about 4 cd/lx.m² per year. Red sheets deteriorate relatively rapidly. Cleaning the sheets can be useful at any time, adding about 10 cd/lx.m², and is particularly effective toward the end of their lives.

The legibility distance of a retroreflective sign (Section 3.2.4) increases logarithmically as its CIL increases. More generally, if a light source has a luminance intensity of LI, the illumination on the reflector will be $b/D2$, where b is a constant and D is the distance between the light and the reflector. This produces a luminance at the reflector of CIL.$A.b/D2$, where A is the area of the reflector. A person standing a distance D_f away (ignoring angular effects) would then perceive an illumination of CIL.$A(b/DD_f)^2$. Reflectorized signs are placed skew to the observer to minimize the amount of unnecessary light reflected back to the driver.

3.1.3.2 Variable Message Signs (VMS)

Part of the "blurring" referred to in Section 3.1.2 is that active systems in which the sign messages may vary have become more common as information technology has improved. *Variable* (or *changeable) message signs* (VMS) were first used in the early 1970s. VMS can have messages added or deleted manually, remotely by a manually initiated command, or remotely and automatically. They are particularly common in situations in which traffic flow varies greatly, for example, in tidal flow, congestion, incident-prone conditions, or where unexpected local weather events can have an impact on driving conditions. Because they can respond to traffic, it may be argued that VMS should not be categorized as totally passive. The most common technologies used for VMS are: (1) flip faces and rotating prisms, containing an entire character, word or message, and (2) a dot matrix, fed by incandescent bulbs, optical fibers, or light-emitting diodes (LED).

LED are based on the ability of some crystals to emit a narrow color band of light when an electric current is passed through them. Initial LED were largely based on gallium arsenide. More recently, LED based on indium and phosphorus have become more common and have been more widely used as the range of available colors has increased. For example, red LED are based on aluminum, indium, gallium, and phosphorus.

The crystals are embedded in solid plastic, which makes the devices very robust. LED typically last at least 10 years, need less maintenance, are more reliable, are easily dimmable, and use less energy than incandescent globes (they can often be supplied by local solar power). Nevertheless, only fiber optics can produce the white preferred for many traffic signs. In addition, LED light intensity (Lay, 1998, Chap. 24) decreases as the temperature rises. In 1998 ITE issued an interim purchase specification for LED signals. It is necessary to check LED colors for people who are color-blind (Lay, 1998, Chap. 16) or wearing sunglasses.

3.2 TRAFFIC SIGN THEORY

3.2.1 General

The basic requirement for a traffic sign is that it be capable of fulfilling an established need for traffic information. This will usually be specified by the relevant traffic authority. Once the need is clearly defined, conveying the required information from signs to road users relies on the use of *legends*, i.e., words and numbers conveying literal messages (e.g., "Keep Left"), or *pictorial elements* such as graphic symbols, shapes, and colors.

The *words* that can be used are discussed in Section 3.2.5 and Section 3.2.6. *Symbols* (Section 3.2.7) may be abstract (e.g., a speed derestriction sign) or indicative (e.g., a cross for a crossroads sign). Common symbolic signs include the class of prohibitory signs (Section 3.1.1), which use a red annulus around the actual symbol and a red diagonal slash through the symbol. The advantages of symbolic signs are their:

- Increased legibility and conspicuity due to the use of larger sign elements (this can double the legibility distance of the sign; see Section 3.2.5).
- Ability to be comprehended rapidly and easily, even when their color is obscured.
- Potential for overcoming problems associated with illiterate drivers.

Some signs are hybrids, e.g., the words "Bus Only" in a green annulus to denote a route reserved for buses. Standard shapes and colors are discussed in Section 3.3.1.

The potential effectiveness of a sign can be checked by considering that its intended user must successfully pass through the following four stages, often as a consequence of a single glance. In these stages, the road user:

- *Detects* the sign. For detection to occur, the sign must be:
  (a1) Visible (Section 3.2.2).
  (a2) Conspicuous (Section 3.2.3).
- *Reads* the sign. For reading to occur, the sign must be legible (Section 3.2.4):
  (b1) At an adequate distance (Section 3.2.5).
  (b2) In the time available (Section 3.2.6).
- *Understands* the sign (Section 3.2.7). For understanding to occur, the sign must be:
  (c1) Comprehensible.
  (c2) Unambiguous.
  (c3) Precise.
- *Acts* on the sign in the intended fashion (Section 3.2.8). For an appropriate action to occur, the information on the sign must be:
  (d1) Credible.
  (d2) Correct.
  (d3) Appropriate.
  (d4) Timely.

The requirements associated with successfully passing these four stages are now each examined in some detail.

3.2.2 Visibility

Requirement (a1). The prime requirement for a sign to be detected is that it be visible, i.e., it can be usefully seen.

3.2.2.1 Geometric Requirements

Requirement (a1). This requirement first requires the sign to be positioned and oriented so that its size and attitude make it geometrically possible for it to be seen by a road user. This can be checked by land survey data. Each jurisdiction will also have local legal and social requirements concerning where and by whom a sign may be placed. In urban areas, it is also necessary to place signs sufficiently high for them to be seen over parked vehicles — a requirement that may counter some requirements needed to achieve headlight illumination (see Section 3.2.2.2). It is also necessary to avoid distraction from adjacent advertising signs.

Other placement issues will be discussed in subsequent sections as various sign requirements are further developed. For example, road safety provisions will require sign supports to be offset at least a minimum distance from the traveled way (perhaps 3 m) and will often require them to be frangible, i.e., to fall on impact while causing only minor damage to the impacting vehicle (Lay, 1998, Chap. 28). A conflicting safety requirement is that the driver's eyes should not be encouraged to stray from observing the way ahead.

However, Section 3.2.3 will show that a driver has maximum search expectation and maximum visual acuity in a visual cone no more than 10° from the driver's line of sight. On high roads, this is often best achieved by overhead signs in which the vertical clearance of about 5 m will often be far less than the horizontal distance, which can easily exceed 10 m across a couple of lanes, a road shoulder, and a clear roadside verge recovery area. However, many drivers are not prepared to use more than a 5° cone for vertical viewing. A 10-m offset and a 10° cone will require a sign to be 57 m away from a driver when sign "reading" is finished. This is called the "cone" distance.

Only one message should be conveyed per sign. It will be seen in Section 3.2.6 and Section 3.8 that a driver's read-and-respond time is about 2.5 s. This means that signs should be separated by at least 70 m on a road where travel speeds average 100 km/h. A common consequence of the "one message per sign" rule and a widespread tendency to solve all traffic problems by erecting signs is that a plethora of signs is proposed (and sometimes installed) along most road lengths. It is therefore necessary to apply rigorously the "one message" rule and the "2.5-s separation rule" and to suggest strongly that issues not accommodated on a traffic sign be dealt with by other means.

3.2.2.2 Optical Requirements

Requirement (a1). This requires the light signal delivered from the sign to the road user to be sufficient to allow the user to receive an adequate visual signal from a sign apart from other signals received from objects near or surrounding the sign. Technically, this ability is measured by the luminance of the sign and its background. To understand this requires some knowledge of lighting theory (Lay, 1998, Chap. 24). However, the subjective equivalent of luminance is brightness (or luminosity).

The *luminous flux* is the light emitted by a lantern or received by a surface such as a sign face. It is measured in *lumen* (lm), which is a power-related unit. The light emitted by a point source (such as a lantern) per solid angle (steradian) is called *luminous intensity* and is a power-density term. Its unit is the *candela* (cd), one of the seven basic units of the SI system. *Luminance* is the luminous intensity per unit area reflected in a given direction from a point on a lit surface. Its units are candela per unit area, i.e., cd/m^2, and it measures the light reflected from a surface. The subjective equivalent of *luminance* is brightness (Lay, 1998, Chap. 24).

Luminance levels range from –6 to –1.5 log cd/m^2 at night, –1.5 to +1.5 log cd/m^2 in twilight, and +1.5 to +6 log cd/m^2 in daylight. Observation shows that the following sign luminances are relevant for a typical road user:

- At least 0.3 cd/m^2 (–0.5 in log units) is needed to ensure that sign color can be detected.
- At least 3 cd/m^2 (+0.5 in log units) is needed to ensure that sign detail can be resolved. This is thus the threshold performance level.
- At least 5 cd/m^2 is needed for new signs to cover performance degradation in service. This is not easy to achieve at night; Lay (1998, Chap. 24) shows that street lighting will supply 2 cd/m^2 or less and vehicle head-lighting 4 cd/m^2 or less.
- At least 30 cd/m^2 (+1.5 in log units) is needed to avoid significant visual degradation.
- At least 40 cd/m^2 is needed for older drivers.
- Many authorities adopt about cd/m^2 as an optimum sign luminance level.
- More than 100 cd/m^2 (+2 in log units) brings little additional visual benefit, probably because of the negating influence of irradiation and glare.
- Values as high as 1.7 kcd/m^2 are needed to enhance conspicuity in brightly lit urban areas, and even higher levels would be needed in strong sunlight.

Visibility is enhanced as the contrast in *luminance* between the sign and its background increases. This is measured by the *luminance contrast ratio* and depends on the internal and external illumination of the sign. For example, the luminance contrast ratio has a number of criterion values based on Chapter 16 of Lay (1998). These are:

- A *threshold* value of 2, at which a sign detail will just be visible. This ability degrades at luminances of 30 cd/m^2 or less and the decrement becomes dramatic below 3 cd/m^2, when the eye becomes very sensitive to light but suffers a corresponding drop in resolving ability (Lay, 1998, Chap. 24; Hills and Freeman, 1970).
- A value of 6 to ensure *useful* visibility.
- An *optimal* value of about 10, which is the level found in the white-on-red design used for "Stop and Give Way" signs.

The eye's sensitivity to contrast decreases as the background luminance increases (Lay, 1998, Chap. 16 and Chap. 24). For *night-time* visibility, the luminance is best achieved by permanent lighting. This may be from special lamps at the front of the sign, by back-lit signs, by surfacings capable of emitting their own lighting, or by motor vehicle headlights.

The luminance of unlit signs depends on retroreflectivity of the sign face, the position and performance of the driver's headlights, and the placement of the sign, which must be low enough to be *illuminated* by the light from the headlights. When illumination by headlights is unlikely, it will be necessary to ensure permanent self-illumination by fixed external illumination or by internal *transillumination* of signs.

Sign shape is best conveyed by illuminating the border rather than the whole sign, which would blur the edges via *halation* (Lay, 1998, Chap. 16). Thus, increasing overall sign luminance can reduce the legibility of a sign.

3.2.3 Conspicuity

Requirement (a2). In addition to being visible, a sign must also be conspicuous; i.e., it must attract the driver's attention with certainty and within a short observation time, regardless of its location relative to the driver's initial line of sight (Cole and Jenkins, 1980). Conspicuity thus deals with sensory effects, that is, with features of a sign that force a viewer to give it perceptual prominence. Conspicuity can be subdivided further into attention conspicuity and search conspicuity.

Attention conspicuity is the ability of the sign to attract attention when the driver is not prepared for its occurrence. The attention conspicuity of a sign is related to its:

- *Luminance and luminance contrast ratio* constitute the probability of detecting a visual signal will increase with an increase in its luminous contrast ratio and, to a lesser extent, with its absolute luminance. The luminances needed for conspicuity are many times greater than those needed for visibility, and values as high as 1.7 kcd/m^2 are needed in brightly lit urban areas, although even this level would not be enough to enhance conspicuity in strong sunlight (Bryant, 1980).
- *Size* is a particularly powerful determinant of conspicuity and means that the eye tends to favor nearby objects that provide a large visual angle (Cole and Jenkins, 1982).
- *Edge contrast and sharpness* relative to its background; large edge contours (or borders) are particularly effective.
- *Color* does not contribute much to the conspicuity of a sign, possibly because the color property does little more than compensate for the luminance loss associated with using color rather than white (Section 3.3.1).
- *Other features* such as surface highlights; shape; graphic boldness; and location relative to the driver's line of sight (Cole and Jenkins, 1982).
- *Relevance* to the driver in his current cognitive state (Lay, 1998, Chap. 16).

A sign with good attention conspicuity will always have good search conspicuity, but the reverse is not always the case (Hughes and Cole, 1984).

Search conspicuity is needed when the object sought lacks attention conspicuity and a visual search is initiated. A driver will take about 300 ms for one *glance* (or *look*) at the scene ahead and about 500 ms to notice a feature in the scene (Lay, 1998, Chap. 16). Thus, visually searching for a sign can consume significant portions of a driver's time.

A measure of search conspicuity is the maximum angle from the eye's line of sight at which an object can be detected with 90% probability within 250 ms (Cole and Jenkins, 1980; 1982). This can be readily translated into traffic engineering practice with respect to sign location and type. For example, common practice is that signs must be placed within 8 to 10° of the line of sight to ensure that they are in a visual zone providing reasonable visual acuity and according to driver expectations (Lay, 1998, Chap. 16). Thus, the search conspicuity measure should be set at 8° or less.

A person's *peripheral vision* usually operates to about 90° on either side of the line of sight. Detection of something moving or visually interesting in the peripheral field will usually lead to an appropriate eye movement. This eye movement process will take at least 500 ms; thus, peripheral signs will not evince a rapid response (Lay, 1998, Chap. 16). Furthermore, the peripheral cone decreases with travel speed to become about 50° at 30 km/h, 40° at 60 km/h, and 20° at 100 km/h.

Traffic control devices are often not particularly conspicuous. In one test series observers located only 50% of the devices for which they were instructed to search, although they did detect 96% of the key regulation signs such as *stop, give way, and speed limit* (Hughes and Cole, 1984). There are also marked differences in the conspicuity of common signs. For example, the give way sign has relatively poor conspicuity and is much less conspicuous than the stop sign. One reason for this is that the word STOP provides a much bolder legend. The give way sign also subtends a smaller visual angle. It has been suggested that its white background results in low contrast relative to a bright sky (Cole and Jenkins, 1982). Other signs that failed to attract attention include parking, street name, and tourist signs (Hughes and Cole, 1984).

Usually only one sign will be noticed in a single glance. Thus, using more than one sign in a driver's effective visual field will result in ineffective signing. Furthermore, giving a driver more observation time will not necessarily increase the number of signs noticed because many drivers will spend the extra time available drawing more information from the first sign noticed (Cole and Jenkins, 1982).

Roadside *advertising* signs cause a small but statistically significant distraction. The general effect is not of great magnitude, at about one crash every 6 years per sign per kilometer per 10,000 veh/d; drivers appear to have defenses against this distraction. However, strongly illuminated signs and signs involving novel, sensuous, flashing, or moving displays of high information content or calling for attention via peripheral vision (Lay, 1998, Chap. 16) should not be located near roadsides. Aesthetic factors will probably play a greater role than safety factors in decisions on advertising sign location.

3.2.4 Detecting Detail

Requirement (b), legibility. In addition to ensuring a sign's visibility, its message must also be made legible; i.e., sufficient detail within the sign must be visible at a given distance (requirement b1) and in a given time (requirement b2). These requirements set the criteria for the physical attributes of the sign's message. They will depend on the capacity of the road user's eyes to resolve the detail in the light signal (Lay, 1998, Chap. 16).

Some signs, such as the *stop and give way* signs, can convey their message by their shape and color alone, with shape the far more important determinant of message legibility. Thus, strong emphasis is placed on the border of such signs. However, if the message within the sign must be read — as with a direction sign — its legibility becomes of paramount importance.

Requirement (b1), legibility distance. The first factor is to determine the critical detail that must be legible; this is usually the smallest detail to be resolved within the sign. Once this is established, the next step is to determine the maximum distance at which this detail can be seen. This is called the *legibility distance**, L_d. The application of the *minimum angle of resolution* data for the human eye shows that the legibility distance for 90% of people with normal vision is 1/0.290 = 3.4 m for every millimeter of detail dimension. Algebraically:

$$W/D \geq 0.000290$$

or

$$D \leq 3500W$$

or

$$L_d = 3500W$$

where W is the stroke width, D is the observation distance, and consistent units are used (Cole and Jacobs, 1978). Results for young people will be about 10% better and for old people about 10% worse than the population average. One corollary to this discussion is that drivers with measurably subnormal visual acuity will have measurably less time in which to read the message on a sign (Lay, 1998, Chap. 16).

Using the preceding data, the legibility distance of a *letter* of the alphabet or other complex symbol would be based primarily on the *stroke width* of the component line elements within the letter or symbol. However, the strokes will have width and length, and this ratio influences the outcome. It is found that the optimal letter legibility is achieved with a stroke length-to-width ratio of between five and ten, with a preference of about six. The exact ratio depends on the shape and style of the letter, the separation from other letters, the shape of the word, the sign colors, and such external factors as background luminance, luminance contrast ratio, and viewing distance.

With a near-optimum stroke length-to-width ratio, a letter or symbol can be recognized as a whole without all its fine detail resolved. This gives a minimum angle of resolution of six times the basic population value of 290 μrad (Lay, 1998, Chap. 16), i.e., 1.75 mrad, for well-designed letters and symbols (Bryant, 1982). The associated legibility distance is 1/0.00175 = 600 mm for every millimeter of letter height, H. For L_d in meters and H in millimeters for the more common lowercase letters, this becomes:

$$L_d = 0.6H_{opt}$$

Many design codes conservatively take $^2/_3$ of this value, i. e., $0.4H_{opt}$.

* Visibility distance is the same as the legibility distance but implies that no detail within the object being detected needs to be observed, i.e., that detail and object are the same.

The spacing of letters is also relevant because *visual interaction* can occur between the contours of the individual letters. However, the effect is usually relatively insignificant, and in practical cases letter spacing can usually be determined on aesthetic considerations (Anderton et al., 1974). Nevertheless, an interletter spacing of about 0.3 times letter height appears optimal for legibility. Letter width is ideally about 0.9 times letter height, so a letter and its surrounding space should occupy about 1.25 times the letter height. Some *numerals* (such as 5 and 8) have the potential to be easily confused, so their shapes must be carefully chosen (Hind et al., 1976).

The fonts used for letters and numerals in traffic signs are therefore very carefully chosen from laboratory, prototype, and in-service field tests to ensure that they provide maximum legibility. Many countries use the letter set originally produced in the U.S. by the Bureau of Public Roads, which is based on a California set. The bureau's Modified E Series set is particularly popular.

Section 3.2.3 mentioned that a large edge contour enhances the conspicuity of a sign. However, visual interaction between this contour and the detail within a sign can cause a drop in legibility. Thus, the need for conspicuity and bold graphics must be balanced against the need for legend legibility. In addition, legibility is best achieved with straight borders. Light lettering on a dark background (positive contrast) is usually more legible with illuminated signs than dark letters on a light background (negative contrast) because the light regions with their greater illumination have less chance to overwhelm their borders by irradiation of the letters (Lay, 1998, Chap. 16). However, in strong and uniform daylight, most observers prefer dark letters on a light background.

It is necessary to provide the sign legend and its background with retroreflectivity in order to convey properly any color-coded message at night. To some extent this will lower the legibility of the sign by lowering the luminance contrast ratio between legend and background.

3.2.5 Detecting Whole Words

The effective *legibility distance*, L_d, is obviously the distance at which the intended message can be read. It is not always necessary to depend on the details of the individual letters seen because legibility is also a function of the shape of the word or symbol and its familiarity to the reader. There is a redundancy in written words, which means that road users can often recognize a word without distinguishing every detail of each letter in the word. This is particularly so for familiar messages. Thus, legibility distance is much greater than would be predicted from knowledge of the stroke widths of the letters and the driver's visual acuity. This effect is much more pronounced if upper-case (i.e., capital) letters are avoided and lower-case letters are used wherever possible. This is because the lower-case letters give words a varying contour, whereas upper-case words are all of a rectangular shape.

Although common experience, supported by test results, is that MESSAGES SOLELY IN UPPER-CASE LETTERS ARE MUCH HARDER TO READ QUICKLY, it is curious to observe how alarm and emergency messages and the newer VMS messages (Section 3.1.3.2) use upper-case letters, presumably on the basis that perceived importance is of higher value than the content of the message.

Clearly, legibility distances are best determined by tests on the specific message or sign in question. Given these qualifications, typical legibility distances in good light are 60 m for a finger-sized detail, 80 m for complex signs, 125 m for legends, and 250 m for symbols and fiber-optic signs (Bryant, 1982; Jacobs et al., 1975). However, these distances can be reduced by such factors as:

- Distraction due to roadside activity.
- Drivers not in a "normal, alert" condition (e.g., affected by alcohol, fatigue, or drugs).
- Low luminance levels at night or due to rain, snow, or fog (Section 3.2.2), dropping legibility distances by up to 50% (Hills, 1972).
- Headlight seeing distances as low as 30 m and rarely above 90 m, although retroreflective sheeting restores much of this night-time loss (Lay, 1998, Chap. 24).
- Dirty and aging signs causing drops of about 30%.

3.2.6 Glance Legibility

Requirement (b2), legibility time. The discussion so far has concerned the distance at which a sign can be read. Another form of legibility relates to reading the sign during quite brief periods of exposure, which is called *glance legibility.* Clearly, the limited time that a driver has to read a sign will restrict the length and complexity of the message that can be extracted. From peripheral vision studies, the maximum legend length, LL, for certain resolution in a single 500-ms glance (Lay, 1998, Chap. 16) is given by:

$$LL = 54W - 0.024D$$

When this is rewritten as:

$$W/D \geq 0.000440 + (LL/54D)$$

it is seen to be much more demanding than the preceding expressions for a single detail.

In a single glance a driver can resolve and read about one new word, or six characters (Cole and Jacobs, 1978). The number of words N that can be read over a longer period T is given approximately by:

$$T = [0.32N - 0.2]\text{ s}$$

which is about three words per second and is consistent with glance times (Jacobs and Cole, 1978a; Lay, 1998, Chap. 16). However, the resolution of a few words may be sufficient to enable the whole message to be "read," particularly if it is a familiar one. Thus, if the sign has a familiar message or a relevant context, then an extra two words could be assumed to be read, i.e., about five words per second. Many design codes assume:

$$T = 0.25N \text{ s}$$

For map signs, a common finding is that:

$$\ln T = 1.043 + 0.054N$$

Short-term memory can retain about seven words or similar chunks of information (Lay, 1998, Chap. 16).

Given that the distance traveled in meters by a vehicle maintaining a steady speed of v km/h over time T s is $vT/3.6$, the distance traveled while N words are read is $Nv/14.4$ m. The legibility distance (Section 3.2.4) must occur when the reading commences and so the sign placement distance, L_p, ahead of this point (Figure 3. 1) and provides the following equation:

$$L_d = 0.6H_{opt} = Nv/14.4 + L_p$$

or

$$H = 0.12Nv + 1.7L_p$$

Equations like this, with minor coefficient changes for safety factors, etc., appear in many design manuals for determining lettering size, H, and thus sign size (see Figure 3.1).

The reading process will be enhanced if as much of the sign as possible is within a driver's field of central retinal vision. Indeed, a driver's zone of maximum visual acuity is usually only a degree or so on either side of the line of sight (Lay, 1998, Chap. 16). Obviously, increasing the size of a sign will improve its legibility distance and give a driver more opportunity to observe it. However, some of the data read at each observation (or glance) will be lost because more of the larger sign will now be seen by peripheral parts of the retina, where acuity drops and contour interaction susceptibility increases (Anderton and Cole, 1982). Large signs can thus be visually inefficient. On the other hand, the reading process will be degraded if the driver's attention is diverted by the demands of driving in heavy or threatening traffic.

If a common driver sign-response time is T_{cog}, then a sign spacing of at least $T_{cog}v/3.6$ is necessary; thus, 60 m would give drivers about 2.2 s between signs when traveling at 100 km/h. Following Section 3.2.2 and the limited reading times available to a driver, only one sign of a particular type should be attached on any single post. The location and use of intersection signs such as stop or give way are discussed in Lay (1998, Chap. 20); some comments on their conspicuity are given in Section 3.2.3.

3.2.7 Understanding the Sign

Requirement (c1), comprehensible. The discussion to date has concerned the sensory perception of a sign. *Comprehension* is stage 3 in the list in Section 3.1 and relates to the driver's response to the sensory perception. The basic process is one

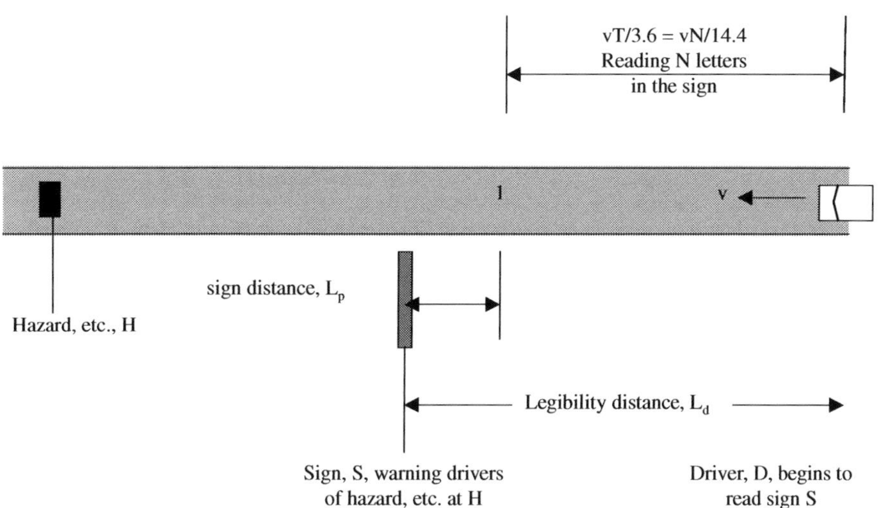

FIGURE 3.1 Sign reading distances.

of pattern recognition. The pattern received must relate to a pattern coded in the driver's memory. Four key questions related to comprehension are whether the perceived message is:

- Meaningful (e.g., to a driver from outside the area).
- Correctly understood.
- Interpreted unambiguously (requirement c2).
- Responded to cognitively in the intended manner.

Requirements (c2 and c3), unambiguous and precise. In assessing whether a sign message will be comprehensible, it is important to understand the:

- Exact nature of the message that the sign is intended to convey.
- Context of its use.
- Population to whom it is directed.

Given the answers to these three points, it is then necessary to ask whether the message will be interpreted unambiguously to produce the intended outcome.

The answer to these questions may involve extensive laboratory and field tests of representative samples of the driving population. The design of and results from these tests must be handled with considerable statistical care. Comprehension can be measured in terms of the speed and the effectiveness of this response. In cases in which an immediate response is not required, the best test is often retention of the message in short-term memory. These issues are particularly important for symbolic signs.

Blank signs may arise with VMS (Section 3.1.3). Surveys suggest that this confuses drivers, so it is probably best to always carry some message on a VMS. Traffic engineers should also view the location of a sign in the context of older drivers with poor memories and longer response times or drivers new to an area without local knowledge.

3.2.8 Acting on the Sign

Requirement (d), correct action. The previous sections have discussed whether a driver could "read" the message on a sign. This section discusses whether the driver will respond to that message. A cognitive desire to act as well as a physical ability to act must be present.

3.2.8.1 Cognitive Response

Requirements (d1–3), credible, correct, and appropriate action. Requirement (c) will have established that the sign message has meaning. A sign message will command cognitive attention so that the reception of its sensory stimulus in the driver's mind will lead to its being given cognitive priority if it has relevance, importance, or novelty. Relevance and importance will usually relate to whether:

- The driver must act on the sign.
- The message will affect the driver's well being.
- Previous signs have been perceived to be useful.

On the other hand, a driver who notices a conspicuous but irrelevant sign will give it little or no cognitive priority, unless it has a degree of novelty. Indeed, considerable evidence indicates that drivers take little notice of many signs (Bryant, 1980). This particularly applies to signs that are obeyed only when risk of prosecution is high. Studies of VMS (Section 3.1.3) giving route diversion data indicate that about a third of the drivers in a position to use the information actually do so.

3.2.8.2 Physical Response

Requirement (d4), timely action. If a driver is to act on the message provided on a sign, the sign must be located sufficiently in advance to allow the driver to properly detect, read, understand, and then respond to its message. Figure 3.1 illustrates how the correct location is chosen, using the case of a sign warning a driver, D, approaching from right to left of a hazard, H, on the road ahead.

Working backwards, the approaching driver must be able to stop before the hazard. The known *vehicle stopping distances* (Lay, 1998, Chap. 19) are therefore used to locate the last point, B, at which the braking process can safely begin upstream of the hazard. The stopping distance for constant deceleration is given by BH in meters, where

$$BH = (v/3.6)^2/2d$$

where v is the vehicle speed in km/h and d is the deceleration in ms^{-2}. Before the vehicle can begin decelerating, the driver must read the sign, decide what to do, do it, and then wait for the vehicle to begin to respond. These times, which will increase

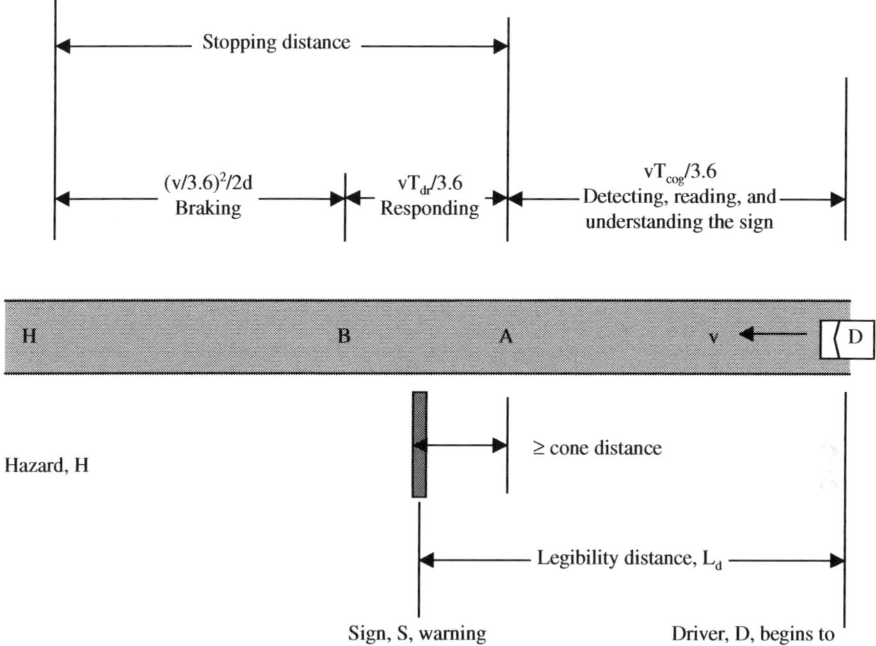

FIGURE 3.2 Sign location assumptions based on permitting driver action to occur. Sequence is to establish H, then A, then D, then S, and then check AS.

with driver age and fatigue, are discussed in some detail in Chapter 16 of Lay (1998). With T_{cog} as the sign scanning, reading, comprehending, and deciding time, and T_{dv} as the driver vehicle-response time in seconds, the equation becomes:

$$DH = DA + AB + BH = (T_{cog} + T_{dv})(v/3.6) + (v/3.6)^2/2d$$

Typically, the vehicle braking response time is 0.4 s and the basic driver response time is 1.7 s (Lay, 1998, Chap. 16). Following Section 3.2.6, a two-word sign might take 0.5 s to read. Recall as a caution that a "new" word could well add another 0.5 s, which is a strong reason for keeping sign messages terse. Thus, 2.6 s is a relevant value of time T for sign calculations and 2.5 s is used in many design codes; AB in Figure 3.2 is the distance travelled in 2.5 s. An increase in luminance also decreases the time a driver will take to react to the sign.

The quantities resulting from such calculations are tabulated for various speeds in Table 3.2, which also lists the *vehicle stopping distances*, that could be achieved if drivers used their full braking capacity. A critical driver value is the deceleration to be used. The measured minimum nonskid braking distances in column 3 are for dry weather and favorable conditions (Samuels and Jarvis, 1978). They imply a friction coefficient, f_l, of about 1.2 (Lay, 1998, Chap. 19). The high decelerations*

* Most drivers consider 4 ms^{-2} to be a maximum (Lay, 1998, Chap. 27).

in column 5 also indicate that the figures in column 3 should not be used for design purposes.

Of course, this is an extreme case because many messages will not require a driver to stop the vehicle. On the other hand, signs requiring lane changing in heavy traffic may require greater distances than the stopping distance. A useful rule of thumb that flows from the discussion in Section 3.2.6 is that a driver will need about a further second of sign reading for each additional decision that the sign may require. Thus, complex tourist signs are best read by drivers at low speeds in roadside areas specifically provided for pulling off the road.

Having established the point *D* distant *DH* from the hazard for a driver who has begun the visual process leading up to acting on the sign message in time to avoid the hazard, the sign is located at *S*, which is the legibility distance L_d back toward the hazard from *D*. There are some constraints on *S*:

- It must be sited prior to the hazard, so as not to confuse the driver.
- It must not be sited so far before the hazard as to be irrelevant and/or to be lost from short-term memory (commonly taken as 15 s of travel — about 400 m at 100 km/h).
- The sign must be big enough for the legibility distance, L_d, to be greater than the cognitive distance, *DA*, by at least the cone distance (Section 3.2.1).

A related dilemma is that, although increasing the legibility distance is achieved by making the sign larger, it cannot be so large that portions of the sign fall outside the driver's 10° cone of high acuity vision (see Figure 3.2).

The associated stopping distances for other trucks are obtained by multiplying the distances in column 8 of Table 3.2 by the factors enclosed in Table 3.3. Column 10 of Table 3.2 lists possible *direction sign* distances, which are consistent with a detecting–reading–comprehending time of 3 s for complex signs (see preceding sections) and a low 60-m legibility distance.

3.3 TRAFFIC SIGN REQUIREMENTS

3.3.1 Shape and Color

The visual significance of sign shape has been discussed in Section 3.2.3 and Section 3.2.5. The actual shapes of signs commonly have the meanings given in Table 3.4. These are generally adopted universally (see Chapter 2). Color on signs commonly has the meanings given in Table 3.5. The specific colors employed for traffic signs are tightly specified. They are described technically by their *chromaticity*, which is based on chromaticity coordinates defined using the *CIE 1931 standard colorimetric system* (CIE, 1978). Each chromaticity is represented by a single point on this planar diagram, which is commonly called the *CIE chromaticity diagram*. Colors need to be used with restraint because they lead to a loss of visibility over *white*, mainly due to a loss of *luminance* (Lay, 1998, Chap. 24). The relative visibilities of some

TABLE 3.2
Stopping and Acting Distances

		Minimum, but Nonskid, Values				Typical Values			
Speed (km/h)	Distance Traveled in 2.5 s (m)	Measured Dry Braking Distance[a] (m)	Measured Stopping Distance [col. 2 + 3] (D_s, m)	Constant Deceleration Implied by col. 4 (ms^{-2})	Time to Stop Using Deceleration in col. 5 (T_s, s)	Car Stopping Distance[b] (m)	Observed Truck Stopping Distance (dry/wet) (m)	Stopping Distance Assumed in Design[c] (m)	Possible Direction Sign Distance[d] (m)
Column Number									
1	2	3	4	5	6	7	8	9	10
40	28	6	33	1.8	10	50	40/60	35	40
60	42	13	54	2.5	11	90	80/120	65	60
90	62	26	88	3.5	12	170	180/250	140	90
100	69	33	102	3.8	12	200	220/300	170	100
110	76	40	116	4.0	12	230	270/360	210	110
120	83	47	130	4.3	12	270	330/420	250	120
130	90	55	145	4.5	12	310	390/480	300	130

[a] These distances will effectively double on unpaved roads. See Lay (1998, Chap. 27).

[b] Lay (1998, Chap. 19): $S = Tv/3.6 + (v/3.6)^2/2f$ with the friction coefficient a conservative design minimum value of $f = 0.3$ (Chap. 12.5.2 and Chap. 19.2.4). This thus gives stopping distances much higher than the values in column 4 for hard braking in good conditions.

[c] From Austroads (1989). Guide to the geometric design of rural roads. NAS-62, Austroads: Sydney.

[d] From NAASRA (1988). Traffic control devices. Guide to traffic engineering practice. Part 8. National Association of State Road Authorities (NAASRA), Sydney.

TABLE 3.3
Associated Stopping Distances for Other Trucks[a]

Situation	Multiplying Factor
Antilock brakes	0.9
Design assumption	1.0
Very good driver or rigid truck	1.1
Bus or empty truck	1.2
Prime mover or articulated truck	1.4
Poor driver	1.5

[a] Obtained by multiplying the distances in column 8 of Table 3.2 by the factors in this table.

TABLE 3.4
Common Meanings of Sign Shapes

Name	Shape	Usage
Octagon (red)	○	Stop
Disc (red)	●	Regulation associated with speed and pedestrians
Equilateral triangle (pointing up, yellow); diamond (yellow)	△; ◆	Warning
Equilateral triangle (pointing down, red)	▽	Give way (or yield)
Rectangle (vertical)	▤	Generally used for regulation
Rectangle (horizontal, green); Rectangle (horizontal, yellow)	□	Guide signs; warning of roadwork
Shield	●	Route marking
Diagonal cross (or crossbuck)	×	Railway level crossings[a]

[a] However, the alternative use of a symbolic steam train symbol (Cole and Jacobs, 1981) has its advantages.

common colors are: white, 1.00; yellow, 0.95; green, 0.7; red, 0.1; blue, 0.05; and violet, 0.0005.

On the other hand, for the same luminance levels, people perceive other colors to be brighter than yellow or white. The retroreflective materials used for traffic signs are discussed in Lay (1998, Chap. 24). The legibility of the individual letters, numerals, and symbols used on signs is discussed in Section 3.2.4 and Section 3.2.6.

3.3.2 Direction Signs

Intersection *direction signs* are guide signs used to: (1) indicate or confirm directions and destinations that may be reached by using one or more of the lanes available to

TABLE 3.5
Meaning of Colors Used on Signs

Color	Usage
Red	Stop, give way, wrong way, and speed reduction signs, etc. — normally means "extreme" hazard or prohibition
Black, white	Regulatory signs and backgrounds and legends on other signs
Yellow	Warning, advisory, and temporary signs —normally used for warning of hazards
Orange	Warning of roadworks
Green	Permissive, guide, and direction signs
Blue	Guide, direction, mandatory, and service signs
Brown	Tourist and recreational information signs

the driver, (2) help locate the intersection, and (3) give some guidance through any channelization (Lay, 1998, Chap. 20).

Intersection direction signs are often supported by *advance direction signs* placed before the intersection, particularly when the complexity of the intersection or the level of the traffic flow requires drivers to be in the correct lane well in advance of the intersection. *Reassurance direction signs* are placed after the intersection. A distinction should also be made between advance direction signs prior to an interchange and the simpler signs needed for at-grade intersections where speeds will be lower and early lane choice will be less critical. Advance direction signs can be made relatively complex because they are scanned rather than read in full; on the other hand, reassurance signs contain minimal information. The two basic forms are:

- The *diagrammatic sign* has destinations positioned on a map-like representation of the intersection and directions represented by the position of the town name on the diagram. Such a sign typically contains two to ten destinations.
- The *stack sign* has destinations in a vertical list (or stack) and directions represented by a small adjacent arrow. The search time for finding a name in a stack will depend on the location of the name and the length of the stack. Typically, the time T needed to search a stack of N names is (Jacobs and Cole, 1978b):

  $$T = [0.25N - 0.17]\text{ s}$$

  which is about 80% of the time given in Section 3.2.6 for reading the words in a line.

  Thus, stacks of up to seven names can be searched in 1.5 s with 95% success. However, design codes often recommend a maximum of three destination names or five lines of information, with no more than two names for each direction of travel.

  For a given sign area, stack signs are superior to diagrammatic signs in terms of legibility because the destination names on stack signs can be made relatively larger. It is commonly believed that diagrammatic signs have a superior ability to convey complex messages. However, Hoffmann and MacDonald (1977) were unable to detect any such superiority in laboratory experiments. In tests on short-term memory retention, no significant difference was found between the two types of signs. Diagrammatic signs were preferable when (Hoffmann and MacDonald, 1980) (1) the necessary direction violated driver expectancies; or (2) a simple graphic design could be found to represent the situation.

Major routes are marked by special signs intended to supplement direction signs. They permit a route to be traversed by following the appropriate numbers and shapes and therefore must usually be used in conjunction with a good map and some advance planning or knowledge.

3.3.3 Safety

There can be no doubt that poor signage can lead to drivers who are distracted, confused, flustered, and/or angry. In these circumstances, they are likely to undertake a maneuver unexpected by other drivers, to accept dangerously small gaps, and to "fail to see" (in a cognitive sense) other road users. Time devoted to scanning the visual scene to retrieve missed information or to correct misunderstood messages may well be time better spent detecting potential conflicts with other vehicles.

In addition, signs may actually create a traffic hazard when struck by errant vehicles. Signs located in an exposed position should use *frangible or breakaway* poles (Lay, 1998, Chap. 28). If it is not possible to avoid placing the sign supports in a potentially hazardous position, they should be isolated from errant traffic by a system of barriers. The placement of sign supports and the design of safety barriers are well documented in modern traffic engineering manuals (e.g., Lay, 1998, Chap. 28) and no justifiable reason exists for a sign being a safety hazard or for critical signs being removed from service. On two-way roads, signs are normally placed only on the side of the carriageway associated with the driver's direction. For safety reasons, signs should be placed in the center median of a divided road only if they have special relevance to traffic in the median lane.

3.4 CONCLUSIONS

The principles of sign design are now well established; high-performance products are widely available and maintenance regimes are well understood and easily implemented. There is no reason that signs should not deliver their intended message clearly and adequately and without creating hazards to road users. Today's road travelers have high expectations and modern signing systems should always meet such expectations.

Realizing that a modern road might cost millions of dollars per kilometer to construct, why not use a small percentage of that cost to provide and maintain signs to ensure that travel on the road is safe and efficient? Finally, it must be noted that signs can form the worst forms of visual clutter, producing an ugly roadside along which a plethora of signs degrades all signs in the minds of road users. Section 3.2.6 discussed the placement of major signs, but at a local level the main factor determining the placement of curbside signs often appears to be the availability of a free meter of curb. Road managers need to be ever-vigilant to ensure that road signs are appropriately designed, placed, and maintained and that only signs critical to the driving task are placed in the driver's normal line of sight.

REFERENCES

Anderton, P.J. and Cole, B.L. (1982). Contour separation and sign legibility. *Austr. Road Res.*, 12(2), 103–109.

Anderton, P.J., Johnston A.W., and Cole, B.L. (1974). The effect of letter spacing on the legibility of direction and information signs. *Austr. Road Res.*, 5(5), 100–102.

Austroads (1989). *Guide to the geometric design of rural roads.* NAS-62, Austroads: Sydney.

Blake, P. (1964). *God's Own Junkyard.* New York: Holt Reinhardt.

Bryant, J.F. (1980). Signs of high brightness. *Proc. 10th ARRB Conf.*, 10(4), 252–262.

Bryant, J.F. (1982). The design of symbolic signs to ensure legibility. *Proc. 11th ARRB Conf.*, 11(5), 161–171.

Cole, B.L. and Jacobs, R.J. (1978). A resolution limited model for the prediction of information retrieval from extended alphanumeric messages. *Proc. 9th ARRB Conf.*, 9(5), 383–389.

Cole, B.L. and Jenkins, S.E. (1980). The nature and measurement of conspicuity. *Proc. 10th ARRB Conf.*, 10(4), 99–107.

Cole, B.L. and Jenkins, S.E. (1982). Conspicuity of traffic control devices. *Austr. Road Res.*, 12(4), 223–238.

CIE (Commission Internationale de L'Eclairage) (1978). Light as a true visual quantity: principles of measurement. CIE Pub. No. 41 TC-1.4. CIE: Paris.

Hills, B.L. (1972). Measurements of the night-time visibility of signs and delineators on an Australian rural road. *Austr. Road Res.*, 4(10), 38–57.

Hills, B.L. and Freeman, K.D. (1970). An evaluation of the luminance contrast requirements of highway signs. *Proc. 5th ARRB Conf.*, 5(3), 57–94.

Hind, P.R., Tritt, B.H., and Hoffmann, E.R. (1976). Effects of level of illumination: stroke width: visual angle and contrast on the legibility of numerals of various fonts. *Proc. 8th ARRB Conf.*, 8(5), Session, 25, 46–55.

Hoffmann, E.R. and MacDonnald, W.A. (1977). A comparison of stack and diagrammatic advance direction signs. *Austr. Road Res.*, 7(4), 21–26.

Hoffmann, E.R. and MacDonnald, W.A. (1980). Short-term memory for stack and diagrammatic advance direction signs. *Proc. 10th ARRB Conf.*, 10(4), 17–22.

Hughes, P.K. and Cole, B.L. (1984). Search and attention conspicuity of road traffic control devices. *Austr. Road Res.*, 14(1), 1–9.

Jacobs, R.J. and Cole, B.L. (1978a). Acquisition of information from alphanumeric road signs. *Proc. 9th ARRB Conf.*, 9(5), 390–395.

Jacobs, R.J. and Cole, B.L. (1978b). Searching vertical stack direction signs. *Proc. 9th ARRB Conf.*, 9(5), 396–400.

Jacobs, R.J., Johnston, A.W., and Cole, B.L. (1975). The visibility of alphabetic and symbolic traffic signs. *Austr. Road Res.*, 5(7), 68–86.

Lay, M.G. (1998). *Handbook of Road Technology.* Vol. 2, 3rd ed., New York: Gordon and Breach.

Macdonald, W.A. and Hoffmann, E.R. (1978). Information coding on turn restriction signs. *Proc. 9th ARRB Conf.*, 9(5), 361–382.

NAASRA (1988). Traffic control devices. Guide to traffic engineering practice. Part 8. National Association of State Road Authorities (NAASRA): Sydney.

Samuels, S.E. and Jarvis, J.R. (1978). Acceleration and deceleration of modern vehicles. *Proc. 9th ARRB Conf.*, 9(5), 254–261.

United Nations (1952). Convention on road signs and signals. Geneva (the protocol was first issued in 1949 and the next convention after 1952 was in Vienna in 1968 — UN Conference on road traffic, Vienna, Final Act and related documents).

4 The Effectiveness of Transport Signs

Cándida Castro, Tim Horberry, and Francisco Tornay

CONTENTS

4.1 Introduction 49
4.1.1 Laboratory vs. Field Studies 50
4.2 Ways to Measure Effectiveness of Traffic Signs 50
4.2.1 Eye Movements/Visual Behavior 51
4.2.2 Memory of Traffic Signs (Recall Method) 52
4.2.3 Driver Responses 53
4.2.4 Naming Signs 53
4.2.5 Accident Rates 53
4.3 Driver Motivation and Awareness with Respect to Signs 53
4.4 Strategies to Enhance Traffic Sign Effectiveness 56
4.4.1 Using New Signs 56
4.4.2 Repeating the Sign 57
4.4.2.1 Implications of Repeating the Sign 59
4.4.3 Presenting Signs on the Same Post 60
4.4.4 Bordering the Sign 62
4.4.5 Reducing Visual Clutter 63
4.5 Conclusions and Practical Implications 65
References 65

4.1 INTRODUCTION

Human error, in its various guises, is the main or a contributory factor in many transport accidents. For certain types of rail–road intersection accidents, the U.K. Department of Transport stated that approximately 15% could be attributed to traffic signs (1993). Indeed, in the U.S. it has been calculated that more than $20 in accident costs could in some circumstances be saved for every dollar spent on roadway safety (figures quoted by Thieman and Avant, 1993). Clearly then, as mentioned in Chapter 3, traffic signs can be critical for traffic safety and considerable cost savings can sometimes be made with their correct use.

0-415-31086-5/04/$0.00+$1.50

Reading traffic sign information correctly is crucial. It helps the transport operator (car driver, pilot, or train driver) to anticipate future situations, make decisions, and start to carry out appropriate motor responses. As was seen in Chapter 2, the effectiveness of transport signs has been well researched over many years in field and laboratory studies. Older examples of this included Moore and Christie (1963); one example of more recent research is Al-Madani (2001, 2002), described later in this book.

4.1.1 Laboratory vs. Field Studies

When someone drives a car, train, aircraft, or truck, many tasks must be performed simultaneously. This is one reason why the simplified laboratory approach to understanding driver behavior has often been criticized (Rockwell, 1966; Milgram, 1970). The main disenchantment focuses on the difficulty of generalizing from simple laboratory tasks to driving in a complex environment. Despite this, many state-of the-art facilities exist around the world to test traffic signs in controlled, yet realistic, conditions. These include:

- England: Institute for Transport Studies, University of Leeds (http://www.its.leeds.ac.uk/); Transport Research Laboratory (http://www.trl.co.uk/).
- France: INRETS (http://www.inrets.fr/index.e.html).
- North America: many excellent laboratories, one of which is the University of Michigan Transportation Research Institute (http://www.umich.edu/~driving/sim.html).
- Australia: Monash University Accident Research Centre (http://www.general.monash.edu.au/muarc/simsite/simhome.html).

On the other hand, several studies comparing drivers' performance obtained parallel results when carrying out field and laboratory research (Lajunen et al., 1996). Likewise, many good laboratory studies have been performed in order to understand how the driver can extract information from traffic signs (Ells and Dewar, 1979; Jacobs et al., 1975; Whitaker and Sommer, 1986; Horberry et al., 1998; Wogalter et al., 1998; Luna–Blanco and Ruiz–Soler, 2001; Drakopoulos and Lyles, 2001). Looking only at warning signs, Wogalter and Laughery (1996) argued that controlled testing of variables such as fonts, colors, symbols, or borders in warning signs can lead to better designed and more effective warnings, which in turn can have positive safety outcomes. This holds especially true for studies undertaking "like for like" comparisons of signs in laboratories and driving simulators in order to judge relative effectiveness of different versions of a sign.

4.2 WAYS TO MEASURE EFFECTIVENESS OF TRAFFIC SIGNS

Even if only driver-centered methods are considered, numerous ways of measuring traffic sign effectiveness exist. These include: recording eye movements, recognition, naming, subjective opinions, recall (tested by questioning drivers about the traffic signs that they have already passed), traffic accidents attributed to poor signing, and getting people to make a conscious search for all traffic signs as they drive. In addition, other methods (that can loosely be described as nondriver centered) include analyzing the wear of signs over several years, the materials of which signs are made, and their implementation costs.

This chapter will focus on driver-centered methods. Most work has been done in the road transport domain; comparatively little work has been undertaken recording behavior of train drivers or aircraft pilots with respect to signing. Although most of this discussion will be based on road transport examples, it is still mainly applicable to other transport domains.

4.2.1 Eye Movements/Visual Behavior

Because operating a vehicle is a task that relies heavily on vision, recording where an individual is looking when driving seems a very attractive technique. A classic example of this is Shinar et al. (1980), who recorded eye movements when subjects were driving on actual roads. However, this general method is not always very useful due to subjects using their peripheral vision to obtain information. Having drivers consciously search for all traffic signs is also a good strategy to explore their capabilities (Johansson and Rumar, 1966) but is not very ecologically valid. Besides, signs can be fixated without processing their messages.

Another classic example is Zwahlen (1981), who carried out a study exploring drivers' eye scanning of warning signs on rural highways. His main results showed that most drivers on average look at warning signs 2.3 times. However, the fixation durations are longer at nighttime and shorten with the distance from which they look at the sign. In addition, he also found significant differences between the fixation times for different types of warning signs. The research in this field is still prolific; a few examples of more recent work follow.

Geoffrey et al. (2001) examined the effects of aging, clutter, and luminance on visual search for traffic signs embedded in digitalized images of driving scenes. They found that errors were more common among elderly subjects and search efficiency declined with increased clutter and with aging. Nevertheless, older adults did not suffer disproportionately as a result of increased clutter relative to the younger ones.

Underwood et al. (2002) also analyzed selective searching while driving. Their results highlighted the role of experience in hazard detection and general surveillance. In another current work, these authors explored a different topic related to the location of traffic signs and the effect of attending to the peripheral world while driving (Crundall et al., 2002).

Another related issue examined by Shinoda et al. (2001) concerned drivers' abilities to detect stop signs visible for short periods of time in a simulated environment. Detection performance was strongly modulated by the instructions and the local visual context. From their results, it can be concluded that visibility of signs requires active search and that the pattern of this visual search is influenced by previous knowledge of the probable structure of the traffic environment.

4.2.2 Memory of Traffic Signs (Recall Method)

Measuring drivers' memory of signs by means of drivers' being stopped on the road and requested to recall a traffic sign which they just passed (Johansson and Rumar, 1966; Johansson and Backlund, 1970; Sanderson, 1974; Aberg, 1981; Drory and Shinar, 1982; Milosevic and Gajic, 1986) is a procedure with certain drawbacks. The main problems are related to: (1) the forgetting processes (because a certain amount of time has passed); (2) the drivers' shock of being stopped, which may influence memory; and (3) the virtual impossibility of conducting such a technique in a transport domain other than road driving. (Imagine the difficulties in stopping a train driver and asking him to report the railway signs he has just passed; see Figure 4.1).

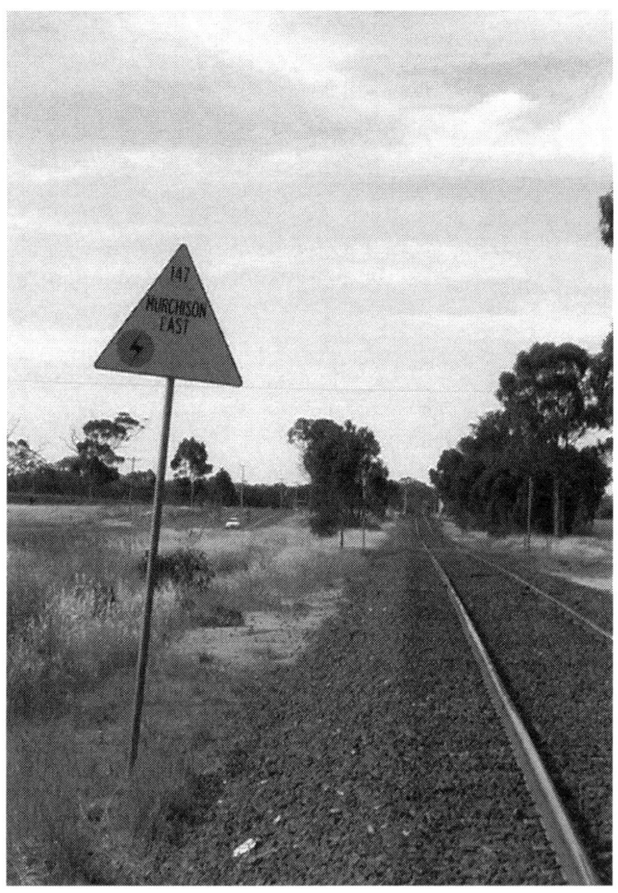

FIGURE 4.1 An example of a railway sign. This sign is difficult to study by a memory test.

4.2.3 Driver Responses

If it is agreed that perhaps the main function of traffic signs (especially warnings signs) is to induce safe behavior on the part of the driver, then obtaining driver behavioral responses to traffic signs (Häkkinen,1965; Sremec, 1973; Summala and Hietamäki, 1984) is an approach with a great deal of face validity. This is especially true when these behavioral methods are combined with other techniques to attempt to uncover some of the psychological processes underpinning such behavior (e.g., memory questions).

Häkkinen's (1965) most important result was that three of four subjects complied with the traffic sign and also recalled it correctly; however, Fisher (1992) found an important inconsistency between drivers' verbal recall and their driving behavior. His procedure consisted of hitchhiking and consequently asking the driver about two previously located road signs. His main finding stressed that memory of signs is generally poor and that they must be studied more thoroughly, not only in terms of recall accuracy but also in terms of their capacity to sensitize the driver to a hazard.

Summala and Hietamäki (1984) evaluated speed adjustments across sign conditions and found significant effects of sign conspicuity and content. Specifically, flashing warning lights were more effective than the same signs without a light and warning signs appeared to be more effective than indication signs.

4.2.4 Naming Signs

Naming traffic signs when experimenters are placed as passengers in the vehicle with the driver was the task used by Johansson and Rumar (1966) and Summala and Näätänan (1974). Sign naming was found to be 97% accurate, despite the fact that this questionable method suffers from low ecological validity. Drivers generally do not drive along naming every sign they see, and the link between the naming and actual safe driving behavior is unproven.

4.2.5 Accident Rates

Accidents are usually multicausal and many transport accidents are difficult to attribute directly to signing shortcomings (a point made in several places in this book). However, it is sometimes possible (for instance, at accident black spots) to install new signs and measure their effects on accident rates. By comparing these data with before-accident rates, it is possible to assess the effectiveness of the sign.

4.3 DRIVER MOTIVATION AND AWARENESS WITH RESPECT TO SIGNS

The majority of the preceding methodologies suggest the general conclusion that the percentage of drivers who correctly recall traffic sign information is small and that drivers are unaware of information presented by most traffic signs (MacDonald and Hoffmann, 1991). Thus, it seems that a sign can sometimes be ignored by drivers (in the road transport environment, at least). It can be argued that the factors influencing driving awareness of signs are deficient motivation, lack of driving experience, and environmental complexity factors. These will be considered in later chapters in more detail and will be briefly reviewed now with respect to traffic sign effectiveness.

First, Näätänen and Summala (1976) suggested that drivers' motivation would be influenced by information redundancy and the presence of incongruent messages. In other words, drivers ignored traffic sign information when it was redundant or not important and when the message was not always followed by the real situation announced. Summala and Hietamäki (1984) reported that the more significant the signs were (e.g., real danger), the greater was the drivers' immediate response to it. "Speed Limit 30," "Children," and "Danger" were the traffic signs that the authors used (from a higher to a lower importance score, according to the authors). In addition, MacDonald and Hoffmann (1991) reported that one of the most important factors influencing the level of reported sign information was the sign's "action potential," e.g., the rated probability of drivers needing to make an overt response related to the sign information. Of course, it is unclear if such arguments would apply in nonroad transport environments in which there is less visual "clutter."

The majority of road signs are redundant and not needed by drivers (Näätänen and Summala, 1976; Luoma, 1986), which easily leads to loss of credibility (MacDonald and Hoffmann, 1991). Many warning signs, especially, lose their relevance when drivers encounter them often without the corresponding source of danger (Fuller, 1984; Summala and Pihlman, 1993). Drory and Shinar (1982) emphasized that (1) the installation of warning signs is abused in that they are often not warranted and (2) under normal daylight conditions, many warning signs are redundant or irrelevant to the driver's perceived needs and the driving task. (They might be useful only under degraded visibility conditions, such as bad weather and nighttime, or for something that cannot be seen, such as a slippery surface.) These authors reported recall levels of between 5% during the day and 10% at night; later they reported recall levels of less than 10% during the day and 16.5% at night (Shinar and Drory, 1983). In addition, Zwahlen (1981) carried out research exploring differences in eye movements when scanning warning signs during the day and night. He found that the fixation durations during nighttime are considerably longer than during daytime. This finding fits in with Drory and Shinar's work (see Figure 4.2).

Second, although Johansson and Backlund (1970) and Aberg (1981) did not find significant differences in recall between experts and beginners, according to other studies, driving experience seems to be a relevant factor that influences drivers' awareness of signs. Häkkinen (1965) reported higher levels of sign recall for inexperienced drivers and young drivers. Milosevic and Gajic (1986) also found greater awareness of signs among young drivers than among professional drivers and more frequent drivers. Furthermore, MacDonald and Hoffmann (1991) reported that inexperienced drivers exhibited a significantly higher level of recalled sign information than experienced drivers. They believed that all drivers tend to ignore all signs with low relevance to an "action potential," and that, as drivers gain experience, signs lose credibility.

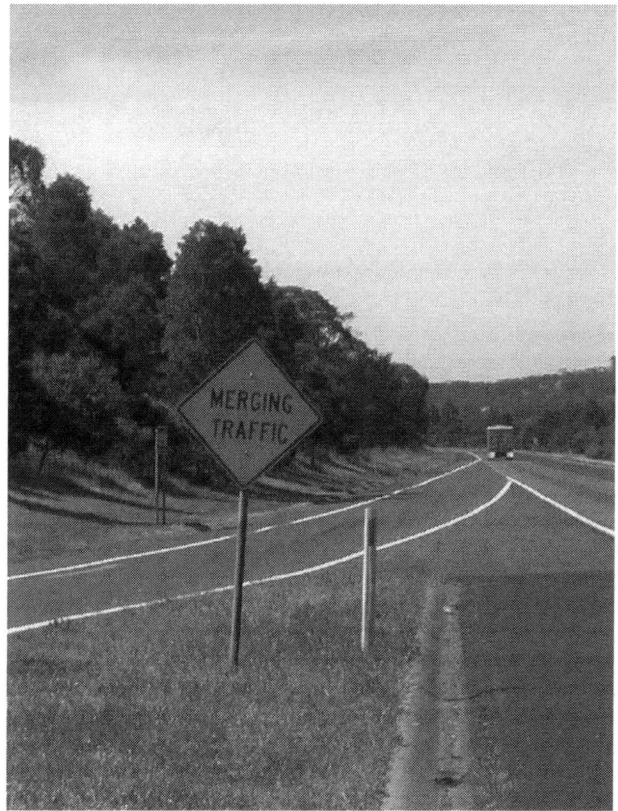

FIGURE 4.2 A sign that might be better recalled at night.

An interesting new line of research involves examining the role of motivation to seek thrills and adventure and avoid boredom for detection of danger on roads (indicated by traffic signs and other traffic devices) or choices made in conditions of road dilemmas (e.g., road-crossing and amber lights) (Rosembloom and Wolf, 2002). The authors found that age and gender differences were a significant factor in relation to sensation seeking.

Finally, other environmental factors could be related to whether a sign is noticed. MacDonald and Hoffmann (1991) listed several of the most important factors. Some of them affect drivers' attentional capacities (such as traffic density, restriction of the view ahead by a crest or curve, and driving speed) and others affect traffic sign conspicuity (such as sign placement, symbolic vs. text sign information, road width, and background complexity). Undertaking similar work in transport domains other than roads would be useful.

A study by Jenkins (1982) demonstrated that two main factors determine the conspicuity of a sign: its size and luminance (related to its contrast with the background). Cooper (1988) also evaluated different ways of increasing traffic sign conspicuity by manipulating different backgrounds (foliage, street scene, trees on skyline) and traffic sign size and color. His results showed that conspicuity can be increased by using larger versions of the signs or fitting the signs with a colored surrounding board. However, he also emphasized that the contrast between the board and the immediate background is a more important factor than sign size.

Furthermore, Cole and Hughes (1990) stated that some physical parameters are determinants of conspicuity, notably eccentricity, background complexity, contrast, color, and boldness of the internal structure of the object. In a domain other than road transport, Young (1991) examined different methods of increasing the noticeability of warning signs. His results showed that pictorials, color, and icons can enhance the noticeability of the signs' warning information.

Concerning the effect of legibility distance, Shinar and Drory (1983) reported that the level of reported sign information was enhanced when the traffic sign was located before rather than after a crest of a hill. They related this result to lack of information about road conditions ahead from nonsign sources. However, MacDonald and Hoffmann (1991) reported other data; they showed that in both conditions the subject's sign information reporting was very low. They believed that the legibility distance is not the major determinant of the reported sign information, even when this distance is limited by an obstruction such as a curve in the road, dusk, or night driving.

4.4 STRATEGIES TO ENHANCE TRAFFIC SIGN EFFECTIVENESS

As suggested in the next chapter by Dr. Terry Lansdown, a "systems" approach to the design and implementation of visual information in the transport environment is needed. New methods to increase traffic sign effectiveness (such as adding more signs) should only be undertaken by examining the influence of such changes on the transport system as a whole. Thus, piecemeal changes need to be undertaken with great care; this caveat should be kept in mind when considering the following options.

4.4.1 USING NEW SIGNS

A strategy to enhance traffic sign effectiveness might be to renew their designs and create new ones. Such newly designed signs usually have positive results, at least during a short period of time. Sanderson (1974) found recall levels varying from 3.5% for a word "Children" warning sign up to 41.2% for an unfamiliar symbolic "Children" sign.

In addition, Summala and Hietamäki (1984) found that a driver would respond equally to signs for "Danger," "Children," and "30 km/h Speed Limit" in terms of taking his foot off the accelerator when first seeing the sign but seemed to control his response to achieve the desired speed better in the "Danger" and "Children" conditions (see Figure 4.3). However, the effect was less significant in all conditions when the signs were accompanied by a flashing light. The new arrangement enhanced the three signs' effectiveness, although the exact mechanism for that improvement is unclear.

FIGURE 4.3 Pictogram (black and yellow) proposed by ANATEEP in order to improve child pedestrian safety. (Reproduced with permission from Transport Scolaires.)

In the area of installing advance warning signs for passive railway crossings, according to Ward and Wilde (1995), results varied in terms of driver behavior. Installing new signs increased drivers' relevant visual search behavior and resulted in speed reductions. However, they did not promote an increase in drivers' stopping or what was categorized as safe behavior. They concluded that advisory signing alone was not usually sufficient to improve crossing safety; therefore, other measures at the site (for example, engineering measures for increasing the visibility of the hazard) were also needed.

When subjects are familiar with a road environment, Ward and Wilde (1995) suggested that the addition of a sign may not result in actual changes of behavior. In a nonroad area, this fact is supported by Goldhaber and DeTurck (1988), who found that using "No Diving" warning signs in a swimming pool did not affect actual diving behavior for swimmers who regularly used the pool, although it did affect the diving behavior of swimmers who were not regulars.

From such research it may perhaps be concluded that newly designed signs can be effective in terms of safe driver behavior. However, their effectiveness declines with time (and increased familiarity). Therefore, the costs of changing to new signs (whether in a road, rail, or other transport environment) might be a serious issue, for which any possible savings through safety improvements or similar changes might be offset by high implementation costs.

4.4.2 Repeating the Sign

One topic often studied in psychology and reading research is semantic and repetition priming in visual word and symbol recognition. The semantic priming effect can be explained through one of the early experiments in the area. Meyer and Schavaneldt (1971) required subjects to press one key if both of two successive visual letter strings were English words or to press a different key otherwise. Unsurprisingly, subjects were faster and more accurate in responding to displays containing two semantically related words (e.g., bread and butter) than to displays with two unrelated words (e.g., doctor and butter). In repetition priming, the prime and the target would be the same word. Though the prime and the target clearly are semantically related, any semantic priming that occurs would be contaminated by other types of priming produced by the shared graphemic and phonemic properties.

Regarding transportation, two facts that influence drivers' interpretation and recognition of the traffic signs should be highlighted. First, when driving (on a road, at least), more than one traffic sign can often be seen in quick succession. Second, the traffic signs may belong to different semantic categories according to the message that they show. The signs may belong to different categories, including: giving orders (e.g., stop), warning (e.g., crossroad), and indicating direction and information (e.g., ring road). So the next step is to ask a question: could repetition or semantic priming effects be found when recognizing information given by traffic signs in real driving environments? In other words, could drivers spend less time and reach a higher accuracy in recognizing traffic signs when two consecutive traffic signs from the same category are presented (e.g., two warning signs) or the same traffic sign is presented twice?

Normally, traffic signs are located in a place corresponding with the message that they convey. In a transport environment such as a road, railway, or airport ramp, drivers meet a sign sequence that they should recognize and interpret. A possibility for increasing the effectiveness of signs that has not received much research attention is simply to repeat the sign at a later point in the transport environment. For example, it seems likely that the effectiveness of a sign warning of a school ahead will be more effective if the sign is placed at several locations before the hazard (school). As yet, however, research examining possible sequences for effective placement of traffic signs has received scant attention.

Castro et al. (Castro, 2000; Castro et al., 2001, 2003, in press, submitted) consistently found the repetition priming effect but only occasionally found the semantic priming effect when using traffic warning signs as stimuli. Another study based on Castro and colleagues' work (Crundall and Underwood, 2001) also reached similar conclusions. Furthermore, at the time of writing (in 2003), work is being undertaken in Australia by Cloete et al. that examines signing sequences for multiple message highway signs, partly based on the previous work by Castro et al.

Before the work of Castro et al., repetition and semantic priming effects had rarely been studied in the context of transport safety. A study carried out by Milosevic and Gajic in 1986 is one of the few studies that have documented the effect of repeated presentation of the same sign. In three experimental situations, they explored driver recall after one of the signs mentioned earlier had been presented once. In the fourth situation, they studied driver recall after the signs "Speed Limit 20 km/hour" and "Road Work in Progress" had been presented simultaneously. The fifth and last experimental situation involved repeated presentation of the same sign: "Speed Limit 20 km/hour." In this experimental situation, the distance between the test and the pretest signs corresponded to 100 m. The authors' sample of stimuli was poor, however; they presented the sign "Speed Limit 20 km/hour" twice, without an appropriate experimental control.

The subjects in their study drove normally along the road where the experimental arrangements had been made. After a bend, the police would stop them and a civilian would ask them about the last sign they had seen. Milosevic and Gajic found levels of recall that ranged from 2 to 20% when the sign was presented once; these levels increased to about 34% when the sign was presented repeatedly. They proved that the repeated presentation of the same sign in a short time interval considerably increased sign registration. The authors explained these results by assuming that, in this case, the drivers had two chances to register the sign in the environment. Therefore, the increase in recall of repeated presentation of the same sign can be explained by the sum of registration probabilities.

One other exception was research carried out by Avant et al. (1996). These authors analyzed the effect of the simultaneous and sequential presentation of two signs that required the driver to carry out the following actions: stop, move right, move left, and slow down. Two signs were always presented, either simultaneously or sequentially. The physical correspondence of the signs, their semantic correspondence, and the use of different signs were also manipulated. They corresponded physically when the same sign was presented with the same format and meaning and corresponded semantically when the same message of the sign was presented, once worded and once symbolically. The signs were different when they did not correspond in format or meaning.

When the participants carried out a same–different categorization task (that is, saying whether the two signs presented were the same or different), they reacted faster and with greater accuracy when the signs were presented sequentially than when they were presented simultaneously. Simultaneous presentation slowed responses and increased errors.

The presentation of physically identical signs produced faster and more accurate responses than presentation of signs with an identical meaning but a different form (a symbolic sign and a worded sign with the same meaning). The authors argued that having to interpret two visual codes (picture and word) with the same meaning slows responses and decreases the accuracy of subjects. When unrelated signs were presented and two semantic codes represented in a symbolic and a worded form had to be processed, decision times increased considerably and accuracy decreased dramatically. Avant et al. concluded by arguing that traffic signs with clearly different formats should be redesigned and that multiple ways of expressing left and right turns, for instance, should be eliminated.

4.4.2.1 Implications of Repeating the Sign

Taken together, the preceding data suggest that the repetition priming effect is valid with warning and indication traffic signs. Given the relevance of both categories, it is pivotal to apply this result to the real driving environment in order to provide information aimed at optimizing drivers' perception of their environments and allowing them to anticipate in the required maneuvers. This can be achieved by presignaling the warning and indication signs considered most important.

In fact, the concept of presignaling has already been successfully used in some transport situations — to indicate motorway exits, for example. An arrow may be used to indicate the lane that the driver must choose to exit the motorway and might inform the driver that this will happen at a certain distance ahead (for example, 1 km). A second sign may inform the driver that the exit is 500 m away. A third sign may indicate that the exit is 100 m. The last sign may indicate the destination with a surrounding border with a different color.

Even though this approach has proved to be useful because it helps the driver anticipate a future maneuver, the data from Castro et al. allow one further step. The repetition priming effect not only can be applied in these cases but also is useful in the case of an immediate maneuver, for instance, the close presignaling of any warning sign ("Roundabout Ahead"). Therefore, the effectiveness of almost any sign can be increased if it is shown repeatedly.

One final issue to note is that the Castro et al. data showed that warning signs are identified more quickly and accurately than indication signs if they are symbolic (not worded) and indication signs are identified more quickly and accurately if they are worded. This, to some extent, mirrors the previous general literature on the topic, which showed the superiority of pictures (among others, Jacobs et al., 1975; Ells and Dewar, 1979; Evans and Ginsburburg, 1985; Owsley and Sloane, 1987; Kline et al., 1990; MacDonald and Hoffmann 1991; Kline et al., 1992; Kline and Fuchs, 1993) and of warning signs (among others, Shinar et al., 1980). This effect may be due to the fact that in interurban settings (in Europe, at least) warning signs tend to be expressed symbolically, whereas indication signs tend to be worded or symbolic.

4.4.3 Presenting Signs on the Same Post

It must also be remembered that, in the Milosevic and Gajic (1986) study mentioned in the previous section, two semantically related traffic signs were presented simultaneously (in the fourth experimental situation). More specifically, two warning signs were shown simultaneously, namely, "Speed Limit 20 km/hour" and "Road Work in Progress." The simultaneous presentation of two related traffic signs did not produce a significantly greater recall or percentage of registration compared to isolated presentation of these signs.

What happens when the signs are displayed in two different sizes? The initial point of the research that tried to answer this question suggested that a large object would delay the perception of a simultaneously presented small object but that a small object would not delay the perception of a simultaneously presented larger object. This effect is known as the *perceptual dominant–subordinate, large–small delay* (Navon, 1977; King, 1990; also see Figure 4.4).

Does a perceptual dominant–subordinate, large–small delay occur in transport signs? Furthermore, do similar dominant–subordinate delays occur in applied settings when one object is less salient because it is in the shade, is blurred, and so on? These possibilities seem to have received little attention. The Manual on Uniform Traffic Control Devices (2003) does not consider the size interaction possibility; nevertheless, it recommends larger signs for greater visibility. Larger objects in road scenes are identified or discriminated more efficiently than small objects (Cole and Jenkins, 1982; Jenkins, 1982; Hughes and Cole, 1986). Similarly, in 1985, Washington, D.C. replaced all of its 50.8-cm stop signs with 76.2-cm stop signs, with no consideration of the effect that the larger stop sign might have on perception of other traffic signs mounted at the same location (Federal Highway Administration, 1989). This situation may be generalized worldwide and to other transport domains in which more than one sign are present.

FIGURE 4.4 (See color insert following page 154.) A possible example of the dominant–subordinate large–small delay.

However, King et al. (1991) unexpectedly found that if the second sign was logically irrelevant, the identification response was not delayed. Furthermore, a small second sign resulted in an almost significantly faster identification response than did having no second sign. The odd results found by King et al. encouraged Castro and Martos (1995) to carry out a study in which they employed one traffic sign or two contiguous traffic signs on a post in a photograph of a road scene. Their results showed that one sign stimulus results in a faster response than two sign stimuli. Large signs produced the fastest reaction and a contiguous second sign resulted in a slower response to the target sign, as predicted in the literature. According to Castro and Martos' data, a general recommendation is that perception is faster when only one sign appears on a post. If this situation cannot be avoided, then they recommend mounting two signs on the same post with the higher-priority one located on the top in a larger size than the other one. Current research (for instance, Luna-Blanco and Ruiz-Soler, 2001) still focuses on this general topic.

4.4.4 Bordering the Sign

Another variable previously found to influence traffic signs' effectiveness is the addition of a border to a sign. Previous research has recommended using a larger border to increase the effectiveness of a sign. For instance, Edworthy and Adams (1996) found that larger borders could increase the perceived urgency of a sign. In the traffic environment, Spijkers (1991) found that creating the background of a sign in a different color from the environment behind it enhances its recognition. For example, sign background colors such as blue (against a cloudless sky), gray (against concrete buildings), or green (against grass) would not be easily recognized. Additionally, several researchers have found that using a colored border on a traffic sign can increase its conspicuity (Cooper, 1988; Thompson-Kuhn et al., 1996) and improve its detection rate by drivers (Dunne and Linfield, 1993; also see Figure 4.5).

A possibly negative effect of using borders was, however, raised by Young (1991). His work on product warning information found that adding a border to a warning sign does not significantly reduce search and recognition time for the warning. Furthermore, he cautioned that the closely fitted border around warning text can reduce its legibility. This is the phenomenon of "contour interaction" (Young, 1991). Young goes on to assert that if the border is far enough from the text, it should improve response times. He does not, however, quantify exactly how far apart the border and words need to be or what the beneficial spatial frequency of the sign should be.

FIGURE 4.5 Traditional borders used in real signs.

To sum up, the study reported previously suggests that the addition of colored borders can have a positive influence on the effectiveness of traffic signs as long as they are not too close to the words (message) contained in the sign.

4.4.5 Reducing Visual Clutter

Along most public roads around some city center railway lines and so on, traffic signs are located near other signs and advertisements. Although a main requirement of any traffic sign is that it be easily noticed, in many actual situations, this requirement is not met. According to Boersema and Zwaga (1990), this lack of conspicuity is caused by several sources: the sign itself (e.g., too small), the location of the sign (e.g., far from the normal line of sight), objects in the vicinity of the sign (e.g., distracting architectural elements or advertisements), the actual layout of the environment (e.g., winding road), and the cognitive state of the user (e.g., nervousness, fatigue, or uncertainty).

A large amount of laboratory and experimental work has been done on searching for targets in cluttered visual fields (Williams, 1966; Treisman and Gelade, 1980; Jenkins and Cole, 1982; Cole and Jenkins, 1984; also see Figure 4.6). However, it is often difficult to generalize such results to complex realistic scenes and practical conditions. Some studies used realistic scenes as stimuli in examining the conspicuity of targets. These included:

- Holahan et al. (1978) — the effect of distracting stimuli in the area around a stop sign.
- Shoptaugh and Whitaker (1984) — reaction time to directional traffic signs embedded in photographed street scenes.
- Hughes and Cole (1986) — subject's detection of objects along streets when driving a car.
- Boersema and Zwaga (1990) — advertising and railway station scenes.

These studies showed that distractions presented in a scene caused a reduction in subjects' visual search performance. These results led to the conclusion that advertisements can decrease the conspicuity of routing signs (which, of course, is partly the advertiser's intention).

Using a more ecologically valid and realistic approach, Castro and Martos (1998) analyzed the effect of advertisements on the conspicuity of traffic signs using real traffic signs and real traffic scenes. In addition, the authors tried to ascertain the color contrast effect between the signs and advertisements. The effect of increasing the number of advertisements produced an increase in reaction time that showed a linear and positive relationship. The color contrast between signs and advertisements also has a significant effect. The lower the color contrast is between sign and advertisement, the slower the reaction time is in identifying the target.

Although these results are hardly surprising, it must be pointed out that traffic signs and advertisements are often placed very closely together, especially in cities and main roads, creating cluttered and distracting road environments. These authors question whether the revenue generated by advertising matches the costs of

(a)

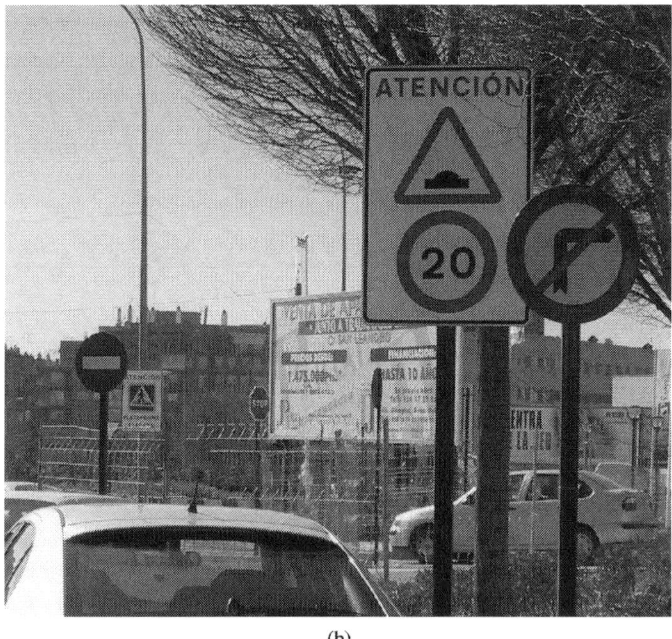

(b)

FIGURE 4.6 (See color insert following page 154.) An example of a cluttered signing environment in Spain.

potentially more frequent accidents (or other negative traffic events) due to drivers' not detecting signs or distraction.

4.5 CONCLUSIONS AND PRACTICAL IMPLICATIONS

A wide variety of methods to evaluate traffic signs may be used. With much of the previous research, it was concluded that the effectiveness of traffic signs (or lack of it) can be enhanced only when perceptual and motivational factors are taken into account. In terms of practical consequences, several methods of improving traffic sign effectiveness were suggested. These were: repeating the sign (and paying greater attention to effective sequences of signs), implementing new signs, adding colored borders to signs, and reducing the "visual clutter" of many transport environments.

Overall, this chapter stressed that controlled testing of signs from the perspective of the human transport operator within a complex sociotechnical system is needed; new sign designs (or redesigns) must take into account humans' strengths and weaknesses as well as the varying demands of driving tasks. This will remain especially important in the future with the changing nature of the road environment, in which optimal Intelligent Transport Systems and Variable Message Signs will be realized only if they are designed and tested from a wide-ranging, driver-centered perspective.

REFERENCES

Anateep (2000). le pictogramme à éclairement est validé. *Transport Scolaires, 128, 14–17* and http://www.anateep.assofr/anateep/pictoigra.htm.

Aberg, L. (1981). The human factors in game–vehicle accidents: a study of drivers' information acquisition. *Stud. Psychol. Upsaliensa*, 6, (Uppsala).

Al-Madani, H. (2001). Prediction of drivers' recognition of posted signs in five Arab countries. *Perceptual Motor Skills*, 92(1), 72–82.

Al-Madani, H. (2002). Assessment of drivers' comprehension of traffic signs based on their traffic, personal and social characteristics. *Transp. Res. Part F. Traffic Psychol. Behav.*, 5(1), 63–76.

Avant, L.L., Thieman, A.A., Zang, A.L., and Hsu, S.-Y. (1996). Memory codes for traffic sign information: visual vs. meaning codes. In A.G. Gale, I.D. Brown, C.M. Haslegrave, S. Moorhead, and S.P. Taylor (Eds.), *Vision in Vehicles, V.* Amsterdam: Elsevier.

Boersema, T. and Zwaga, H.J.G. (1990). Searching for routing signs in public buildings: the distracting effect of advertisements. In D. Brogan (Ed.), *Visual Search*, 151–157. London: Taylor & Francis.

Castro, C. (2000). Acquisition of information while driving. Unpublished Ph.D. thesis, University of Granada, Spain.

Castro, C. and Martos, F.J. (1995). Effect of the size, number and position of traffic signs mounted on the same post. *Sci. Contrib. General Psychol.: Perceiving Imaging Space*, 13, 89–100.

Castro, C. and Martos, F.J. (1998). Effect of background complexity in perception of traffic signs: the distracting effect of advertisements in the proximity of the sign. *Gen. Psychol.**, 143–153.

Castro, C., Horberry, T., and Gale, A. (in 2004). The effect of semantic and repetition priming on traffic sign recognition. In A.G. Gale et al. (Eds.), *Vision in Vehicles, VIII*. Derby: Vision in Vehicles Press.

Castro, C., Horberry, T., Tornay, F.J., and Martos, F.J. (2001). Eficacia de las señales de tráfico: Factores que influencian su percepción. *Boletín Psicología*, 73, 57–80.

Castro, C., Horberry, T., Tornay, F.J., Martínez, C., and Martos, F.J. (2003). Efectos de facilitación de repetición y semántica en el reconocimiento de señales de indicación y peligro. *Cognitiva*, 15 (1), 19–23.

Cloete, S. (2003). Personal communication.

Cole, B.L. and Hughes, P.K. (1990). Drivers don't search: they just notice. In D. Brogan (Ed.), *Visual Search*. London: Taylor & Francis.

Cole, B.L. and Jenkins, S.E. (1982). Conspicuity of traffic control devices. *Austr. Road Res.*, 12, 221–238.

Cole, B.L. and Jenkins, S.E. (1984). The effect of variability of background elements on the conspicuity of objects. *Vision Res.*, 24, 261–270.

Cooper, B.R. (1988). A comparison of different ways of increasing traffic sign conspicuity. Research Report, 157. Crowthorne: Transport and Road Research Laboratory.

Crundall, D. and Underwood, G. (2001). The priming function of road signs. *Transp. Res. Part F: Traffic Psychol. Behav.*, 4(3), 187–200.

Crundall, D., Underwood, G., and Chapman, P. (2002). Attending to the peripheral world while driving. *Appl. Cognit. Psychol.*, 16(4), 459–475.

Department of Transport (1993). Progress to reduce bridge bashing. HMSO: London.

Drakopoulos, A. and Lyles, R.W. (2001). An evaluation of age effects on driver comprehension of flashing traffic signal indications using multivariate multiple response analysis of variance models. *J. Saf. Res.*, 32(1), 85–116.

Drory, A. and Shinar, D. (1982). The effect of roadway environment and fatigue on sign perception. *J. Saf. Res.*, 21, 25–32.

Dunne, M.C.M. and Linfield, P.B. (1993). Driver detection of local and non-primary destinations on direction signs. In A.G. Gale, I.B. Brown, C.M. Haslegrave, H.M. Kruysse, and S.P. Taylor (Eds.), *Vision in Vehicles, IV.* Amsterdam: Elsevier.

Edworthy, J. and Adams, A. (1996). *Warning Design: A Research Prospective*. London: Taylor & Francis.

Ells, J.G. and Dewar, R.E. (1979). Rapid comprehension of verbal and symbolic traffic sign messages. *Hum. Factors*, 21(2), 161–168.

Evans, D.W. and Ginsburburg, A.P. (1985). Contrast sensitivity predicts age-related differences in highway-signs discriminability. *Hum. Factors*, 27, 637–642.

Federal Highway Administration (1989). The federal highway administration's action plan for older persons. Washington, D.C.: U.S. Department of Transportation.

Federal Highway Administration (2003). Manual on uniform traffic control devices (MUTCD). http://mutcd.fhwa.dot.gov. and http://mutcd.fhwa.dot.gov/kno-overview.htm. Retrieved January 21, 2004.

Fisher, J. (1992). Testing the effect of road traffic signs' information value on driver behavior. *Hum. Factors*, 34(2), 231–237.

Fuller, R. (1984). A conceptualisation on driving behavior as threat avoidance. *Ergonomics*, 27, 1139–1155.

Geoffrey, H, Scialfa, C.T., Caird, J.K. and Graw, T. (2001). Visual search for traffic signs: the effects of clutter, luminance and aging. *Hum. Factors*, 43(2),194–207.

Goldhaber, G.M. and DeTurck, M.A. (1988). Effectiveness of warning signs: familiarity effects. *Forensic Rep.*, 1(4), 281–301.

Häkkinen, S. (1965). Perception of highway traffic signs. Report 1. Helsinki, TAIJA.

Holahan, C.J., Culler, R.E., and Wilcox, B.L. (1978). Effects of visual distraction on reaction time in a simulated traffic environment. *Hum. Factors*, 20, 409–413.

Horberry, T.J. (1998). Bridge strike reduction: the design and evaluation of visual warnings. Unpublished Ph.D. thesis, University of Derby, U.K.

Horberry, T.J., Halliday, M., Gale, A.G., and Miles, J. (1998). Road signs and markings for railway bridges: development and evaluation. In A.G. Gale, I.D. Brown, C.M. Haslegrave, and S.P. Taylor (Eds.), *Vision in Vehicles, VI*. Amsterdam: Elsevier.

Hughes, P.K. and Cole, B.L. (1986). What attracts attention when driving?. *Ergonomics*, 29(3), 377–391.

Jacobs, R.J., Johnston, A.W., and Cole, B.L. (1975). The visibility of alphabetic and symbolic traffic signs. *Austr. Road Res.*, 5(7), 68–73.

Jenkins, S.E. (1982). Consideration of the effects of background on sign conspicuity. *Proceedings of the 11th Australian Road Research Board Conference, 11(5), 182–205.* Melbourne: ARRB.

Jenkins, S.E. and Cole, B.L. (1982). The effect of the density of background elements on the conspicuity of objects. *Vision Res.*, 22, 1241–1252.

Johansson, G. and Backlund, F. (1970). Drivers and road signs. *Ergonomics*, 13, 749–759.

Johansson, G. and Rumar, K. (1966). Drivers and road signs. *Ergonomics*, 9, 57–62.

King, D.L. (1990). A large rectangle delays the perception of a separate small rectangle. *Perception Psychophys.*, 47, 353–383.

King, D.L., Sneed, D.C., and Schwab, R.N. (1991). The effect of one and two traffic signs on two measures of perceptual performance. *Ergonomics*, 34, 1289–1300.

Kline, D.W. and Fuchs, P. (1993). The visibility of symbolic highway signs can be increased among drivers of all ages. *Hum. Factors*, 35(1), 25–34.

Kline, D.W., Kline, T.J.B., Fozard, J.L., Kosnik, W., Schieber, F., and Sekuler, R. (1992). Vision, aging and driving: the problems of older drivers. *J. Gerontol.: Psychol. Sci.*, 47, 27–34.

Kline, T.J.B., Ghali, L.M., Kline, D., and Brown, S. (1990). Visibility distance of highway signs among young, middle-aged and older observers: icons are better than text. *Hum. Factors*, 35(5), 609–619.

Lajunen, T., Hakkarainen, P., and Summala, H. (1996). The ergonomics of road signs: explicit and embedded speed limits. *Ergonomics*, 39(8), 1069–1083.

Luna-Blanco, R. and Ruiz-Soler, M. (2001). Within factors involved in the vertical signalling perception: holistic processing vs. analytical. *Psicothema*, 13(1), 141–146.

Luoma, J. (1986). The acquisition of visual information by the driver: interaction of relevant and irrelevant information, Report 32. Helsinki: The Central Organisation for Traffic Safety.

MacDonald, W.A. and Hoffmann, E.R. (1991). Drivers' awareness of traffic sign information. *Ergonomics*, 34(5), 585–612.

Meyer, D.E. and Schavaneldt, R.W. (1971). Facilitation in recognising pairs of words: Evidence of a dependence between retrieval operations. *J. Exp. Psychol.*, 90, 227–234.

Milgram, S. (1970). The experience of living in cities. *Science*, 167, 1461–1468.

Milosevic, S. and Gajic, R. (1986). Presentation factors and driver characteristics affecting road-sign registration. *Ergonomics*, 29, 325–335.

Moore, R.L. and Christie, A.W. (1963). Research on traffic signs. Road research laboratory. *Proc. Eng. Traffic Conf.*, 113–122.

Näätänen R. and Summala, H. (1976). *Road-User Behavior and Traffic Accidents*. Amsterdam: Elsevier.

Navon, D. (1977). Forest before trees: the precedence of global features in visual perception. *Cognit. Psychol.*, 9(3), 353–383.

Owsley, C.E. and Sloane, M.E. (1987). Contrast sensitivity, acuity and the perception of "real word" targets. *Br. J. Ophthalmol.*, 71, 791–796.

Rockwell, T.H. (1966). Skilled judgment and information acquisition in driving. In T.W. Forbes (Ed.), *Human Factors in Highway Safety Research.* New York: Wiley.

Rosembloom, T. and Wolf, Y. (2002). Sensation seeking and detection of risky road signals: a developmental perspective. *Accident Anal. Prev.*, 34(5), 569–580.

Sanderson, J.E. (1974). Driver recall of roadside signs. Traffic Research Report, 1. Ministry of Transport. New Zealand: Wellington.

Shinar, D. and Drory, A. (1983). Sign registration in daytime and night-time driving. *Hum. Factors*, 25, 117–122.

Shinar, D., Rockwell, T.H., and Malecki, J.A. (1980). The effects of changes in driver perception on rural curve negotiation. *Ergonomics*, 23, 263–275.

Shinoda, H., Hayhoe, M., and Shrivastava, A. (2001). What controls attention in natural environments? *Vision Res.*, 41(25–26), 3535–3545.

Shoptaugh, C.F. and Whitaker, L.A. (1984). Verbal response times to directional traffic signs embedded in photographic street scenes. *Hum. Factors*, 26, 235–244.

Spijkers, W. (1991). The recognition of traffic signs under "natural" conditions. In A.G. Gale, I.B. Brown, C.M. Haslegrave, I. Moorhead, and S.P. Taylor (Eds.), *Vision in Vehicles, III.* Amsterdam: Elsevier.

Sremec, B. (1973). Perception of a traffic sign and driving speed. In *First International Conference on Driver Behavior*, Zurich.

Summala, H. and Hietamäki, J. (1984). Drivers' immediate responses to traffic signs. *Ergonomics*, 27, 205–216.

Summala, H. and Näätänan, R. (1974). Perception of highway signs and motivation. *J. Saf. Res.*, 6, 150–154.

Summala, H. and Pihlman, M. (1993). Activating a safety message from truck drivers' memory: an experiment in a work zone. *Saf. Sci.*, 16, 675–687.

Thieman, A.A. and Avant, L.L. (1993). Traffic sign meaning: designer intent vs. user perception. In A.G. Gale, I.B. Brown, C.M. Haslegrave, H.M. Kruysse, and S.P. Taylor (Eds.), *Vision in Vehicles, IV.* Amsterdam: Elsevier.

Thompson-Kuhn, B., Garvey, P.M., and Pietrucha, M.T. (1996). Visibility factors for on-premise advertising signs. *Proc. Int. Conf. Traffic Transp. Psychol.*, Spain, Valencia, 1996, 164–165.

Treisman, A.M. and Gelade, G. (1980). A future integration theory of attention. *Cognit. Psychol.*, 12, 97–136.

Underwood, G., Crundall, D., and Chapman, P. (2002). Selective searching while driving: the role of experience in hazard detection and general surveillance. *Ergonomics*, 45(1), 1–12.

Ward, N.J. and Wilde, G.J.S. (1995). Field observation of advance warning/advisory signage for passive railway crossings with restricted lateral sightline visibility: an experimental investigation. *Accident Anal. Prev.*, 27(2), 185–197.

Whitaker, L.A. and Sommer, R. (1986). Perception of traffic guidance signs conflicting symbolic and direction information. *Ergonomics*, 29, 699–711.

Williams, L.G. (1966). The effect of target specification on objects fixated during visual search. *Perception Psychophys.*, 1, 315–318.

Wogalter, M.S. and Laughery, K.R. (1996). Warning! Sign and label effectiveness. *Curr. Directions Psychol. Sci.*, 5(2), 33–37.

Wogalter, M.S., Kalsher, M.J., Frederick, L.J., Magurno, A.B., and Brester, B.M. (1998). Hazard level perceptions of warning components and configurations. *Int. J. Cognit. Ergonomics*, 2(1–2), 123–143.

Young, S.L. (1991). Increasing the noticeability of warnings: effects of pictorial, color, signal icon and border. *Proc. Hum. Factor Soc. 35th Annu. Meet.* 580–584 San Francisco: Human Factors Society.

Zwahlen, H.T. (1981). Driver eye scanning of warning signs on rural highways. *Proc. Hum. Factors Soc. 25th Annu. Meet.* 33–37 Rochester, N.Y.: Human Factors Society, Santa Monica.

5 Considerations in Evaluation and Design of Roadway Signage from the Perspective of Driver Attentional Allocation

Terry C. Lansdown

CONTENTS

5.1 Introduction 71
5.2 Costs and Benefits of Traffic Signs 72
5.3 Functional Road Signs 73
5.4 Traffic Signs as Communication Systems 73
5.5 Effectiveness of Signage 74
5.6 Constraints on Traffic Signs 75
5.6.1 The Environment 75
5.6.2 Different Needs 75
5.6.3 Variable Message Signs: In-Vehicle Issues 76
5.6.4 Information Selection: Traffic Sign Messages 76
5.6.5 Visual Demands of the Journey 76
5.7 Brief Guidelines for Signage Design to Meet Human Needs 77
5.8 Does the Current Traffic Sign System Work? 78
5.8.1 A Question of Evaluation 79
5.9 Conclusions and Practical Implications 80
References 80

5.1 INTRODUCTION

This chapter primarily focuses on the design and suitability of roadway signage. Sign text is considered from the perspective of the visual demands imposed upon the driver and measures that may be undertaken to mediate these to reasonable levels. Much of the text is also applicable to rail, maritime, and pedestrian information

0-415-31086-5/04/$0.00+$1.50

needs. Signage is considered to encompass all information sources available to the driver. This broad definition is adopted from the standpoint that roadway, traffic, or nontransport-related information may now be presented to the driver using traditionally adopted mechanisms, e.g., road markings and signs, or via more esoteric vectors, e.g., dynamic text to speech roadway information. To consider the primary and most conventional method, road signs provide the driver with basic navigational, safety, and roadway information. They are familiar methods to convey such data and include strong contextual cues to support decoding the information they provide — for example, red circles used to indicate mandatory signs on U.K. roads.

5.2 COSTS AND BENEFITS OF TRAFFIC SIGNS

Road signs have several inherent limitations. Chiefly, these include:

- Frequent use of text to convey meaning (restricting understanding to readers of the language).
- A limited time period available for users to extract information.
- Statutory requirements regarding placement.
- Nonstandardization of signage (on an international level); for example, the various shapes and coding of "give way" signs.

By contrast, road signs have inherent benefits in that, at a national level and therefore appropriate to most users, road signage is predominantly standardized. Road signs are relatively cheap to produce and install and are generally effective in provision of information without undue attentional distraction because of good use of typographical features, visual images, and display properties, e.g., reflective materials and good contrasts. The problematic elements of a driver's interaction with signage are frequently associated with the interaction between the driver and his vehicle, other road users, and the weather and geographic components of the driving environment (Connoly, 1968; Dewar, 1988; Verwey, 1993).

If one takes a historical perspective on the value of road signs, one may consider the role of signs during World War II in Europe: road signs were specifically removed to confuse enemy spies. This strategy nicely indicates the changing value of a road sign based upon the cognitive map of the user. The local driver with geographic knowledge does not require detail regarding navigational directions; thus, consideration must be given to the differing needs of a variety of users. The driver unfamiliar with local geography requires quite different information from the regular user traveling the same route to work each day. It is not sufficient to suggest that the familiar route user no longer requires road signage, although its role is clearly much reduced. This issue becomes more salient when considering more dynamic road information, for example, congestion information. Variable Message Signs offer flexible and specific information to the driver, but their effectiveness has been shown to be largely dependent on the driver's trust in the information presented. This trust will be influenced by the currency of the data, their accuracy, and, fundamentally, their effectiveness for the user (Hanowski et al., 1994; Janssen and van der Horst, 1991; Kinghorn et al., 1994).

Signage is clearly a primary mechanism adopted to support transfer of information about the travel network to the driver, captain, or pedestrian. However, use of signage should be considered with respect to the human's capabilities and habits. As will be seen later in this book, Al-Madani and Al-Janahi (2002) report in an automotive context that only 56% of road signs are comprehended. In support of these data, evidence from an occupational setting suggests that when a warning sign is experienced, just over 80% of users will notice the sign; over 60% will read it; and only 30% will comply with it (HSE, 1999).

5.3 FUNCTIONAL ROAD SIGNS

Roadway signage should be designed to recognize and address in its specification: (1) the correct conspicuity of the sign and (2) the importance of its information; for example, is a secondary reinforcing sign required for important information in recognition of the human's limited attentional capabilities (as has been mentioned earlier in this book)? During a prolonged drive, the ability to extract information effectively is also reduced, as has been shown in many vigilance tasks (Craig, 1991); this further lowers the likelihood of signage detection. Many road networks do employ redundant signage, e.g., speed limit indications. However, a question that may be posed is whether this redundancy has been sufficiently considered with respect to the salience of other road information. To illustrate, a driver who has missed a motorway turning because the junction information was obscured behind a large vehicle may consider the signage redundancy to be insufficient. However, such an analysis should be undertaken with reference to the needs of the various stakeholder groups; for example, one important group, road network managers, may feel additional investment in redundant signage not warranted for navigational data, based on financial grounds.

5.4 TRAFFIC SIGNS AS COMMUNICATION SYSTEMS

Vertical signs provide a simple and straightforward method to communicate with the transport user. They are perhaps the most familiar sign experienced. However, evidence suggests (Ogden et al., 1990) that many individuals experience difficulty in extracting information from vertical signs. Two specific areas of difficulty that warrant further exploration are age and attentional demand.

Older drivers have been shown to experience driving difficulties as a consequence of age-related physiological and cognitive changes. As shown later in this book, numerous studies have indicated a decline in visual performance with age (Kline, 1994; Owsley, 1993; Poynter, 1988; Shinar and Schieber, 1991), more specifically, problems with visual acuity, spatial and temporal contrast, and disability glare. Behavioral modification associated with age has been reported to compensate for increased difficulties experienced when driving. For example, older drivers have been shown to avoid fast road types and busy travel times (Burns, 1997). It seems likely that the age-related decline in visual performance may contribute to changes in travel behavior associated with age; furthermore, such choices may be reasonably expected to apply in nonroad settings.

Visual attentional demand has become a significant concern for the driver in recent years for several reasons, most notably, increases in the availability of in-vehicle technology with the potential to distract. These distractions need to be considered within the context of the driver's capabilities and the environmental demands. Signage available to the driver may be considered to be roadway presented (most typically) or in-vehicle (using travel information and communication systems (TICS)). Roadway-presented signage includes: traffic lights, dynamic and Variable Message Signs, horizontal signs (e.g., markings directly on the road) or vertical signs (ideally presented perpendicular to the driver), and gantry-presented signage.

Typically, considered in isolation, roadway-presented signs do not impose an unreasonable visual demand. However, a substantive hazard from such information sources may occur when they are experienced in combination with other information, e.g., flashing lights or lane markings, with resultant potential for incorrect lane selection or, worse, a collision with other road users. In-vehicle signage may be presented using conventional instrumentation or in-vehicle displays, or via head-up displays (HUD). As will be seen later, applications include dynamic navigation systems, congestion warning, road sign echoing, and mobile-office services like e-mail and Web browsing. The introduction of TICS technology into the vehicle has made the distinction between roadway signage and other, more general information available to the driver difficult to make. Furthermore, the benefits and hazards associated with the use of such technology require careful consideration. For example, in-vehicle echoing of a road sign may be considered highly beneficial if the driver has missed the roadside information, but it could overload the attentional capacity of a driver already experiencing difficulties from a busy roadway environment.

For designers of the transportation system, interesting and careful choices need to be made regarding how and where to encourage the operator to focus his limited attention. To illustrate, head-up displays offer interesting design challenges. The information is frequently presented on the road a short distance in front of the vehicle and is optically focused to infinity to remove potential reaccommodation times from changing focal length to resolve the image. Such a display has potential to benefit the driver by reducing the time to extract information from a display in comparison to an in-vehicle image, but it may obscure or confuse vertical roadway signage. Research suggests that HUD may result in more rapid and accurate detection of information presented than via conventional means (Grant et al., 1995).

5.5 EFFECTIVENESS OF SIGNAGE

Traffic signs' effectiveness will largely be dependent on the user's characteristics — for example, the driver's needs and expectations when interacting with the road network; whether he is active or passive with respect to the driven route; his degree of familiarity with the route; and his state of urgency, fatigue, or relaxation during vehicle use. These transient factors will be compounded by the more stable individual differences. Consider the interference effects of visually distracting signs for the nontarget user. What are the implications of a failure to transfer information, a wrong turn, or missing a "Give Way" sign? Sign design should support anticipation of the network, reduce uncertainty, and, ideally, modify behavior to improve safety.

As mentioned in several parts of this book, signage design is traditionally evaluated by means of legibility and speed and accuracy of comprehension (Allen et al., 1994; Dewar and Ellis, 1994; Dewar et al., 1976). However, a systems approach is advocated to consider the design of signage as one element of the operator, environment, and journey system. The relevance of a sign design to each part of the system needs to be considered.

5.6 CONSTRAINTS ON TRAFFIC SIGNS

5.6.1 The Environment

The infrastructure imposes particular constraints, for example, on selection of roadway signage. The geometry of the three-lane road vs. a single-lane road will unavoidably change the eccentricity of a sign and its trajectory relative to the driver. Statutory regulations regarding placement and position of mandatory, advisory, and navigational information have the consequence of variable, and sometimes visually "busy," road environments. Thus, it is inappropriate to apply a "one size fits all" approach to presentation of roadway signage. The motorway network adopts well-researched methods, including physically larger signs and unique color themes, to aid users and consider specific characteristics of this road type, e.g., higher speeds. However, the city center and, particularly, roadway interchanges among urban, motorway, and rural road types remain problematic for many drivers from an information-processing perspective.

5.6.2 Different Needs

Information needs from signage change during the course of a journey. For example, on unfamiliar journeys a driver initially requires general navigational information regarding the broad geographic area; as the journey nears its end, he needs finer resolution information, e.g., street names, familiar landmarks, and personally meaningful navigational cues. Similarly, information needs change as a consequence of the time of day and direction of the journey. If one drives east in Europe in the morning, the sun creates a disability glare potentially obscuring road signage. Similarly, dawn, dusk, and nighttime driving requires more use of the monochromatic "rods" in the retina than the "cone" type of photoreceptors; accident statistics suggest that driving at these times is more challenging. This is not to suggest that journeys at such times should not be undertaken or that using such routes should be avoided. However, it should be recognized that the vehicle user's needs vary in very quantifiable ways and may be addressed to a greater or lesser extent in the design and positioning of roadway signage.

Practical difficulties with individual differences, operating environments, and journeys, like those outlined earlier, have resulted in confusing roadways and excessive information on much signage, as can be witnessed by drivers' frequently missing signs and making wrong turns. This remains true for the rail environment, also; for example, different speed signage is in common use at the same time for the various train types in the U.K. Beware of quick fixes like additional signs, or bigger and bolder signs, without careful consideration of the broader impact on the vehicle system. The context of the operating environment, journey, and driver will influence proper selection and design of signs.

5.6.3 Variable Message Signs: In-Vehicle Issues

As will be seen later in this book, advanced in-vehicle technologies offer some promise to contend with variations in roadway signage presentation. In-vehicle "echoing" of external road signs enables consistent and familiar information presentation, in the language of the user, with the familiar signage of the home nation. However, substantial usability questions remain, such as how to deal with multiple signs, how to deal with spatial and temporal distributions of road signs, and how to manage driver workload and ensure that the in-vehicle system will not increase the attentional demands on the driver. A fundamental audit, evaluation, and, if necessary, redesign of each and every road element to which the driver is exposed are required to facilitate appropriate information acquisition for the driver and therefore maximize opportunities for safety. Clearly, using an appropriate risk assessment could prioritize such a task.

5.6.4 Information Selection: Traffic Sign Messages

The significance of information presented via signage should be explicitly linked to a hierarchy of priorities. Safety should be the primary concern, with regulatory compliance and travel efficiency high on this list. Design of a sign should meet the conspicuity needs of the driver, e.g., high object conspicuity for speed limit information and search conspicuity for navigational information (i.e., signage containing only information relevant to a subsection of the driving population) (Martens, 2000). Aside from this, nontravel-related information, e.g., advertising, should be recognized as low-priority information and, consequently, constrained in its attention-demanding capacity appropriately, as discussed later in this chapter and elsewhere in this book.

5.6.5 Visual Demands of the Journey

An audit for visual suitability should be undertaken to determine the potential hazards from visually overloading signage. The audit should explicitly consider roadway, environmental, and individual factors contributing to the visual demand experienced by the individual.

The roadway features of the route undertaken will be of substantial interest to readers of this text but would also be applicable to a rail, pedestrian, or waterway network. Relevant considerations include the visual angle of signage from the focus of expansion, the eccentricity of significant information from the direction of travel, the conspicuity (Cole and Hughes, 1984) of the visual image, and the optical flow (Lishman, 1981) or speed of information exposure. Additional factors include the complexity of junctions, volume of traffic, predictability of that traffic, and appropriate behaviors associated with the mode of transport. New developments include, for example, the use of rotating advertising space when positioned at a location where it will distract some drivers at the very point at which they should be fixating on the roadway features (e.g., a sharp or blind bend of a complex junction). Quantification of the point-to-point visual demands of the roadway would ensure that such advertising features be placed only in suitable locations so that undue visual or cognitive demands are not experienced by the operator.

Environmental attentional demands primarily encompass weather-related factors, for example, rain or cold conditions, that will have an impact on the driver's regulation of the rate at which he can attend, assimilate, and act upon cues from the available visual environment. It is pertinent to consider the significance of obscuration of signage by weather conditions; it may be that the information is not of a critical nature (priority elsewhere) and, therefore, it is acceptable to employ a lower set of priorities in placement and selection of appropriate signage.

Evaluation of signage requires an assessment of suitability for purpose. Consideration of the findings from a visual environment audit and the rationalization of these with respect to the needs of the stakeholders are necessary. For example, in some complex city road networks, signage requirements may easily overload the driver, so these must be prioritized to reduce the visual demands to an acceptable level. Several strategies are available to support this redesign, from removal of excessive signage to changes in the emphasis, syntax, or graphic elements used to provide conspicuity. The significant feature of this process must be the specific consideration of the driver's needs when interacting with the junction at a fundamental travel level and, beyond that, the additional capacity available to support "value-added" visual information acquisition. These issues are considered in more detail next.

5.7 BRIEF GUIDELINES FOR SIGNAGE DESIGN TO MEET HUMAN NEEDS

The redesign of signage to support appropriate attentional allocation is a huge topic, so a basic overview is provided below to support the key principles. These include: position, typographic features, color, and reflective/refractive properties. Manipulation of these features will change the emphasis of particular sign elements and the consequent priority the driver will allocate. For example, in general, larger signs will be interpreted as more important than smaller ones.

Use of typographic features encompasses the arrangement and appearance of fonts used in signage. Typically, Helvetica or Helvetica variants are adopted for road signage throughout much of Europe. It would seem reasonable to expect that serif fonts would have better legibility than sans-serif fonts on the basis that the text shape presents additional information to the reader. However, the research evidence is unclear with respect to which font type offers better legibility and readability. The Euroface font (Bügleichenhaus, 1999) was developed for use as a European standard font for road signs. Research evidence suggests that the typeface is 42% more legible than Helvetica when viewed at speeds greater than 80 km/h.

Drivers have strong stereotypes associated with particular colors; for example, traffic light signage reinforces "red equals stop, green equals go" preconceptions. Other signage extensively exploits these stereotypes to convey the significance of particular sign types, i.e., advisory and mandatory indicators. It is effective to use color as a secondary or supporting design feature. Relevant considerations include use of a limited palette only to reduce display confusion and avoidance of color combinations likely to have a color-blindness implication, most notably red–green combinations that affect 10% of males. The traffic light provides an interesting example of the use of color in combination with positional cues so that the red–green color-blind individual may safely interact with signage with an acquired knowledge of the spatial location of the stop and go cues.

The use of reflective/refractive materials for road signs has undoubtedly improved their legibility during dawn, dusk, and nighttime driving. However, the reflective properties may become problematic for some users when "blooming" occurs. Blooming refers to the driver's experience of reduced legibility when the text of a road sign loses shape as a consequence of the light brightly reflected from the vehicle's headlights by the sign. This phenomenon may become more problematic when considering interaction with modern vehicles with HID (high-intensity discharge, predominately Xenon) headlights with higher luminous intensities than traditional halogen bulbs. Such difficulties will be further exacerbated for older adults (over 55 years) as a consequence of the general decline in visual capability. However, it is a consideration of the contribution of the various design elements of the sign within the context of the road network that information presentation within the network may be revised and improved.

5.8 DOES THE CURRENT TRAFFIC SIGN SYSTEM WORK?

The volume of traffic successfully negotiating road networks is a testament to the effectiveness and design efficiency of the various component parts of roadway signage. However, objective criteria for the determination of acceptable visual and cognitive demands on the driver remain an elusive research goal. Several approaches offer promise to define reasonable information-processing limitations for road users; these may be broadly considered as assessment environments and assessment methodologies.

Assessment environments that support definition of visual demand limits include the private test track, vehicle simulator, and primary tracking task. Each environment offers a unique position on a range with ecological validity at one extreme and repeatability at the other. Assessment methods employed may include mental workload assessment, situational awareness (or SA), eye glance measurement, and visual occlusion. ISO work has been undertaken to support these areas and standards exist (e.g., ISO, 2002) or are in progress (ISO, 2001) to find definitions and methods for using these approaches.

5.8.1 A Question of Evaluation

Evaluation of roadway signalization requires an understanding of the visual information acquisition strategies of the driver. Moreover, objective determination of the functional effectiveness of the driver during interactions with signage is required; i.e., did he obtain the information needed to make an appropriate decision? The physiological structures of the eye, i.e., foveal and parafoveal vision, encourage the driver to develop an attentional spotlight (Humphreys and Bruce, 1991), which by definition requires shifting attention from one area to another. This attentional allocation has been shown to be motivated by a desire for "reduction of uncertainty" (Wierwille, 1993). It has been argued that the reduction of uncertainty is governed by the differing objectives of the driver and influenced by the context in which information extraction is taking place (Lansdown, 1996). The relevance of these attentional allocation strategies to signalization is widespread. For example, an overly complex sign may demand visual attention by the driver at the expense of a pedestrian or vehicle.

Four research approaches that quantify the impact of visual demands of road signage on the driver may be employed to functionally evaluate the effectiveness of road signage.

- Eye movement analysis may be undertaken using several technologies, from manual video-tape transcription (Lansdown, 1996) to fully computerized data analysis. Analysis of eye movements is therefore an expensive process in terms of initial equipment expenditure or labor costs for videotape transcription.
- Verbal protocols (Bainbridge, 1981) are used to ask drivers to report their thoughts on a moment-to-moment basis and thus may be employed to determine the conspicuity of a particular sign in context with its surroundings.
- Tachistoscopic presentation of road signage may be employed to determine the duration required to extract information for road signs and therefore is useful in evaluation of different display options and consideration of the absolute time required to comprehend signage. However, it cannot consider the context for which the sign is intended.
- Situational awareness (SA) has been defined as "… the perception of the elements in the environment within a volume of space and time, the comprehension of their meaning, the projection of their status into the near future and the prediction of how various actions will affect the fulfillment of one's goals …" (Endsley, 1995). SA can be effective in determining whether a sign has been noticed and, if so, to what degree the information was comprehended.

These methods may be adopted to determine objective measures of the visual demands imposed by roadway signage. It is suggested that development of a harmonious roadway environment would be enhanced by efforts to quantify, assess, and modify roadway signage from the perspective of the impact it will have on the driver's interactions with the road environment.

5.9 CONCLUSIONS AND PRACTICAL IMPLICATIONS

This chapter has stepped back from the specifics of signage and argued for a more fundamental consideration of its design. The approach advocated is to determine the suitability of the existing operating setting, vehicle user, and environmental system and then design or revise signage to meet the information-processing capabilities of the user. Thus, from the theoretical standpoint of understanding of the operator's attentional allocation, it is suggested that signage will be better able to support regulatory, advisory, navigational, and value-added information needs if designed in context with the needs and capabilities of the vehicle user.

REFERENCES

Al-Madani, H. and Al-Janahi, A.R. (2002). Assessment of drivers' comprehension of traffic signs based on their traffic, personal and social characteristics. *Transp. Res. Part F*, 5, 361–374.

Allen, R.W., Parseghian, Z., and Rosenthal, T.J. (1994). Simulator evaluation of road signs and signals. Santa Monica, CA: Human Factors and Ergonomics Society.

Bainbridge, L. (1981). Verbal reports as evidence of the process operator's knowledge. In E.H. Mamdani and B.R. Gaines (Eds.), *Fuzzy Reasoning and Its Applications*, pp. 343–368. Condon: Academic Press.

Bügleichenhaus, E. (1999). The manual on uniform traffic control devices and type safety. European Committee for Uniformity of Type Design and Type Safety.

Burns, P.C. (1997). Navigation and the ageing driver. Unpublished doctoral thesis, Loughborough University, Loughborough.

Cole, B.L. and Hughes, P.K. (1984). A field trial of attention and search conspicuity. *Hum. Factors*, 26(3), 299–313.

Connoly, P.L. (1968). Visual considerations: man, the vehicle and the highway. *Highway Res. News*, 30, 71–74.

Craig, A. (1991). Vigilance and monitoring for multiple signals. In D.L. Damos (Ed.), *Multiple-Task Performance*, 1st ed., p. 153. London: Taylor & Francis.

Dewar, R. and Ellis, J. (1994). The design and evaluation of traffic signs. Paper presented at the 12th Triennial Congress on the International Ergonomics Association, Toronto, Canada, August 15-19.

Dewar, R., Ellis, J., and Mundy, G. (1976). Reaction time as an index of traffic sign perception. *Hum. Factors*, 18(4), 381–392.

Dewar, R.E. (1988). In-vehicle information and driver overload. *Int. J. Vehicle Design*, 9(4/5), 557–564.

Endsley, M. (1995). Towards a theory of situation awareness in dynamic systems. *Hum. Factors*, 37(1), 32–64.

Grant, B.S., Kiefer, R.J., and Wierwille, W.W. (1995). Drivers' detection and identification of head-up vs. head-down tell tale warnings in automobiles. Paper presented at the Proceedings of Human Factors and Ergonomics Society 39th Annual Meeting, San Diego, California.

Hanowski, R.J., Kantowitz, S.C., and Kantowitz, B.H. (1994). Driver acceptance of unreliable route guidance information. Santa Monica, CA: Human Factors and Ergonomics Society.

HSE (1999). Reducing error and influencing behaviour (HSG48, ISBN: 0717624528): HSE Books.

Humphreys, G.W. and Bruce, V. (1991). Visual cognition: computational, experimental and neuropsychological perspectives. LEA.

ISO (2001). ISO/TC22/SC13/WG8 N322: Road vehicles. Ergonomic aspects of transport information and control systems. Maximum allowable visual distraction for in-vehicle information and communication systems. Draft work item.

ISO (2002). ISO 15007-1 and 2: Road vehicles. Measurement of driver visual behaviour with respect to transport information and control systems.

Janssen, W. and van der Horst, R. (1991). *Descriptive Information in Variable Route Guidance Messages*. Piscataway, NJ: IEEE.

Kinghorn, R.A., Bittner, A.C., and Kantowitz, B.H. (1994). Identification of desired system features in an advanced traveller information system. Paper presented at the Proceedings of the Human Factors and Ergonomics Society 38th Annual Meeting, Nashville, TN.

Kline, D.W. (1994). Optimizing the visibility of displays for older observers. *Exp. Aging Res.*, 20(1), 11–23.

Lansdown, T.C. (1996). Visual demand and the introduction of advanced driver information systems into road vehicles. Unpublished doctoral thesis, Loughborough University, LE11 3TU, U.K., Loughborough.

Lishman, J.R. (1981). Vision and the optic flow field. *Nature*, 293, 263–265.

Martens, M.H. (2000). Assessing road sign perception: a methodological review. *Transp. Hum. Factors*, 2(4), 347–357.

Ogden, M.A., Womack, K.N., and Mounce, J.M. (1990). Motorist comprehension of signing applied in urban arterial work zones. In T.R. Board (Ed.), *Transportation Research Record 1281*: *Human Factors and Safety Research Related to Highway Design and Operation* 1990 (Vol. TRR 1281, pp. 127–135). Washington D.C.: Transportation Research Board.

Owsley, C. (1993). Assessing visual function in the older driver. *Clin. Geriatr. Med.*, 9(2), 389–401.

Poynter, D. (1988). The effects of aging on perception of visual displays: Society of Automotive Engineers (SAE), Report number, 881754.

Shinar, D. and Schieber, F. (1991). Visual requirements for safety and mobility of older drivers. *Hum. Factors*, 33(5), 507–519.

Verwey, W.B. (1993). How can we prevent overload of the driver? In B. Peacock and W. Karowowski (Eds.), *Automotive Ergonomics*, pp. 235–244. Washington, D.C.: Taylor & Francis.

Wierwille, W.A. (1993). Visual and manual demands of in-car controls and displays. In B. Peacock and W. Karwowski (Eds.), *Automotive Ergonomics*, pp. 299–320. Washington, D.C.: Taylor & Francis.

6 Railway Signage

Alexander Borodin

CONTENTS

6.1 Introduction 83
6.2 Signage 84
6.3 Signage Taxonomy 84
6.3.1 Safety-Critical Signals 84
6.3.1.1 Mechanical Signals 84
6.3.1.2 Colored Light Signals 85
6.3.1.3 Fixed Board Signals 86
6.3.1.4 Speed Boards 88
6.3.1.5 Advisory (Informational) 90
6.3.1.6 Passenger Information Signs (Static and Dynamic) 90
6.4 A Word about SPAD 91
6.5 The Future of Railway Signage 92
6.6 Conclusions 93
Further Reading 93
References 94

6.1 INTRODUCTION

Railways were the first form of mechanized transport. While the first trains ran on tracks at relatively low speeds and were limited in number to one train on a track, obviously no need existed, at that time, to present information to the train driver. There were no other trains to run into and no way for the vehicles to exceed the maximum cornering speeds for the track. Unfortunately, with an increase in rail vehicle traffic, engine power, speed, and network complexity, accidents began occurring; this led to the development of the first safety rules, procedures, and signage.

Collisions led to attempts to provide more effective systems of safe working to ensure train separation. An example of this process was the Armagh accident of 1889 that led to industry-wide adoption of mechanically interlocked signaling systems and safe working procedures based on confirming the absence of any axles on a block section before allowing any other train to proceed onto that section (Faith, 2000). At the same time, derailments at points and on curves due to speeding led to development of signage to indicate the lay of point switches and the maximum allowable speeds on particular sections of track. This developmental process is still occurring and will continue to occur as long as railways exist.

0-415-31086-5/04/$0.00+$1.50

This chapter presents an overview of current signage practice and philosophy in the railway industry. The reader should note that the world has thousands of different railways, each with a multitude of unique operational problems that need to be addressed; therefore, it is impossible to cover all of the different signage types. Also included in this chapter is a section on the most recent developments in information delivery and the direction that modern railways are taking in terms of signage.

6.2 SIGNAGE

As seen throughout this book, all signage exists to deliver information in some form. Railway signage exists primarily to inform train drivers and other operational staff of two types of information:

- *Safety-critical information* **must** be acted upon and is encapsulated in the rules of the railway. Examples of signs that display safety-critical information are signal aspects, speed boards, point indicators, and stop boards. Lack of compliance to safety-critical information incurs a safety risk; therefore, it is common practice for railways to enact procedures to monitor and reduce any lack of compliance to safety-critical information.
- *Operational advice* assists train drivers and staff to perform their operational work more effectively but does not impact the safe working of trains if it is ignored. Examples of operational advice signage include kilometer markers and signal nameplates.

6.3 SIGNAGE TAXONOMY

This section pictorially illustrates some of the basic signage types, using examples from several railways. Note that signage designed to achieve the same ends may vary radically in size, shape, color, and context. This is due to the difference in signaling philosophies between railways; indeed, it is possible that certain signs extant on one network may be rare, or even superfluous, on another network.

6.3.1 Safety-Critical Signals

The purpose of signals is to ensure the safe separation of trains. These signs dynamically indicate the status of the section of track ahead of the signal. In the simplest terms, trains are shown a STOP indication if the section ahead is occupied and a PROCEED indication if the section ahead is free.

6.3.1.1 Mechanical Signals

Mechanical signals were the earliest form of railway signals. They consist of a movable arm or disc, the position of which indicates a simple PROCEED or STOP. A mechanical semaphore signal (Figure 6.1) indicates a STOP aspect if the semaphore arm is horizontal and a PROCEED aspect if it is in any other position (generally 45°). Semaphore signals were usually fitted with color lights for

FIGURE 6.1 (See color insert following page 154.) Basic lower-quadrant mechanical semaphore signal showing the STOP and PROCEED aspects, respectively.

interpretation at night. Although generally considered obsolete, mechanical semaphore and disc signals are still used and maintained by many railways worldwide.

6.3.1.2 Colored Light Signals

As incandescent bulb technology became brighter and more reliable, it became possible to remove the mechanical arm and introduce colored light signals that were visible during daylight conditions. These signals are the current norm for most railways and have since benefited from the introduction of super-bright LED technology, which provides excellent readability in even the brightest daylight conditions and minimizes the possibility of phantom signal aspects (i.e., sunlight on the signal lens giving the false impression that the lens is lit).

Colored light signals vary in shape and meaning among railways. The simplest system in use follows British Rail principles of *route signaling* and may consist of two-, three-, and four-aspect signal heads. Other signaling regimes include *speed signaling* and various home-grown combinations of speed and route signaling. Unlike traffic lights on roads, colored light signals provide a higher degree of information about future conditions using a system called *aspect sequencing*. In British and QR (Queensland Rail) use, this provides the train driver with information about the state of the next signal (i.e., the signal ahead). In basic route signaling, the aspects have the following meanings:

- *Red* — STOP. A train is on the next section.
- *Single yellow* — PROCEED WITH CAUTION. Next signal is at STOP.
- *Double yellow* — PROCEED WITH CAUTION. Next signal is at CAUTION.
- *Green* — PROCEED. Next signal is at PROCEED.

The variations on this include the use of flashing yellow and flashing green, but these are specialized cases.

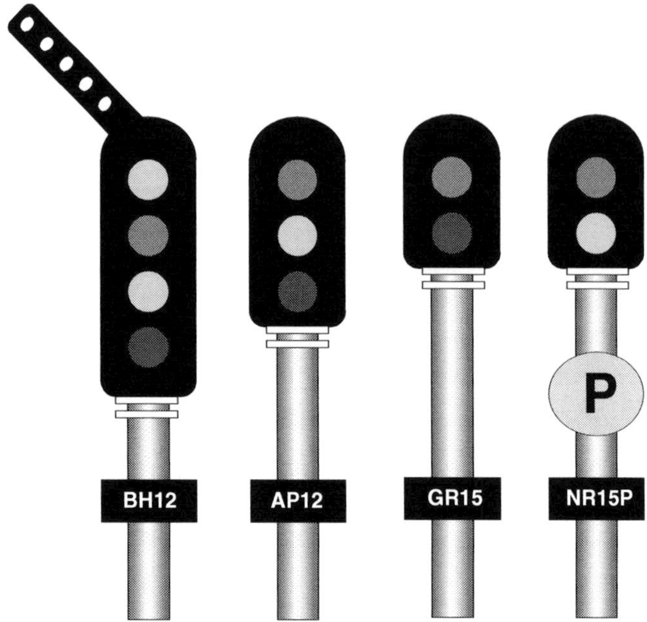

FIGURE 6.2 (See color insert following page 154.) Examples of basic colored light signals (following QR route-signaling standards based on British standards). From left to right: four-aspect signal with left turn-out junction route indicator attached; three-aspect signal; two-aspect signal; and two-aspect signal repeater.

For this style of signaling, a junction route indicator (JRI) is used (see the four-aspect signal in Figure 6.2) to show whether the route is cleared for a junction. For multiple junctions, multiple JRI can be used. This advises the train driver to slow down for a junction to a safe speed. In addition to the preceding, signal repeaters are used where visibility of a signal is obscured. These show whether the signal being repeated is at STOP or PROCEED only and are not, strictly speaking, signals in their own right.

In speed signaling, the colors have different meanings corresponding to the speed that the train is allowed to travel. JRI are not used in speed signaling because the speed aspect is all that is required for the train driver to drive a train safely.

6.3.1.3 Fixed Board Signals

Fixed boards are treated as signals if they display a STOP aspect. Examples of these boards are STOP boards (Figure 6.3), LIMIT OF SHUNT boards (Figure 6.4), BLOCK LIMIT boards, and YARD LIMIT boards.

Certain systems of safe working use marker boards to indicate the beginning/end of a block section. In particular, the train order system of safe working uses yard limit boards to show the beginning and end of a station yard into which a train may have written authority to enter (Figure 6.5). The QR DTC (direct traffic control) system is similar but uses block limit boards, which can be placed anywhere in a

FIGURE 6.3 Fixed STOP board with a red circle. All trains must not pass without authority from train controller.

FIGURE 6.4 Limit of shunt board. Trains may shunt freely up to this board but are not permitted to pass.

FIGURE 6.5 Yard limit board used in train order territory to show the start/end of a station yard.

section and allow for many more trains to fit on a track section than can be achieved in simple train order working (Figure 6.6).

In QR practice, DTC block limit boards and yard limit boards are treated as signals when used in conjunction with a written authority issued by train control.

FIGURE 6.6 DTC block limit board used to show the limit of a block section (QR only).

6.3.1.4 Speed Boards

Speed is limited on railways in much the same way as on roads — by using signage. The difference, however, is that railway signage can often afford to be minimal because train drivers must be route competent.

Permanent speed boards. Australia's National Codes of Practice for the Defined Interstate Network define permanent speed boards as being simple black-and-white reflective boards with numbers affixed to them (Figure 6.7).

Temporary speed restriction boards. Often temporary speed restrictions (TSRs) need to be initiated in order to protect areas of track that are undergoing maintenance or are subject to changed conditions (Figure 6.8). The National Codes of Practice for Australia use signs to enact a TSR. Although all speed boards must be complied with, this is especially true in the case of TSRs because they indicate changed conditions or the presence of work crews on the track.

Dynamic speed indicators. Dynamic speed indicators (DSIs) are used in route-signaled railways as a means of reminding train drivers of reduced speeds through the various branches of a turnout (i.e., set of points). These indicators are generally a "theater" style dot matrix of LED or optic fibers and show a number corresponding to the permissible speed through a turnout. These are a comparatively recent

FIGURE 6.7 Permanent speed boards can indicate a single speed in kilometers per hour (left) or two speeds — the top for passenger trains and the lower for freight trains (right).

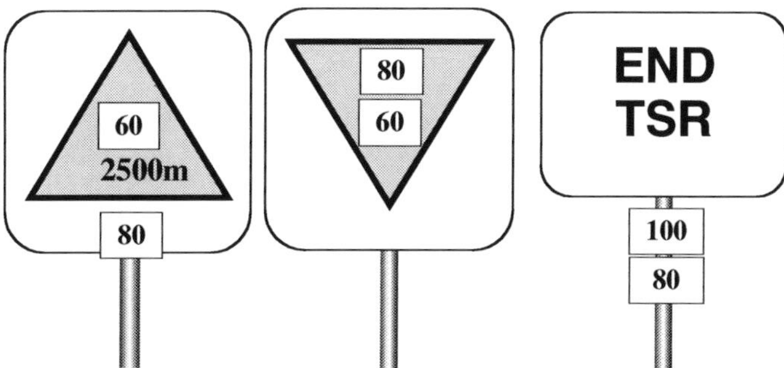

FIGURE 6.8 Temporary speed restriction (TSR) boards are used to indicate changed speed limits. They are used in sequence. The meanings are as follows: the 2500-m TSR warning board (left) is placed 2500 m ahead of the TSR; the start TSR board (center) is placed 50 m before the start of the TSR; and the end TSR (right) board is placed 50 m after the end of the TSR. For all of these signs, different speeds can be included for different classes of trains (per permanent speed boards).

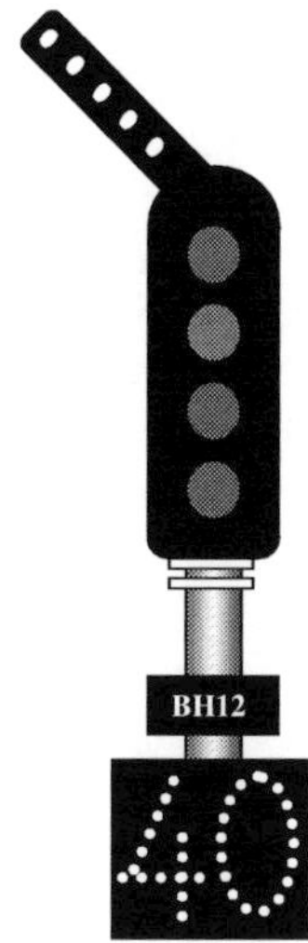

FIGURE 6.9 (See color insert following page 154.) Dynamic speed indicator (DSI) on a four-aspect route signal. In this example, the DSI is lit, showing a 40-km/h speed limit. The DSI is used only if the junction route indicator (JRI) "rabbit-ear" is lit. This allows a train to proceed through a turnout at 40 km/h when, normally, the speed may have been assumed to be 25 km/h.

innovation in railway signaling. An example is shown in Figure 6.9. Various railways use a number of variations on the DSI concept. The most common variation is to omit the least significant digit (i.e., the zero) on the end of the speed. Thus, a "6" indicates "60" km/h.

6.3.1.5 Advisory (Informational)

Advisory signage is used to show nonsafety-critical information for train drivers, track workers, operational staff, and passengers. This information assists people in locating and identifying places and trains.

Signal nameplates. Every signal is identified by a nameplate. Every railway has different conventions for labeling — for example, using the kilometer/mileage of the signal, using the interlocking mnemonic with an associated number, or using a combination of both. Examples of QR style signal nameplates can be seen in Figure 6.2.

Location markers (e.g., kilometer markers). Location markers are used simply to show distance along the track. In Australia, these are kilometer posts and are specified in the National Codes of Practice (see Figure 6.10).

6.3.1.6 Passenger Information Signs (Static and Dynamic)

Passenger information signage is an area in which much development has been focused in recent years, particularly in terms of dynamic (electronic) information display. Many passenger railways, particularly those with complex networks and/or complex timetables, require the ability to communicate to passengers the departure time and platform of particular trains as well as their stopping patterns. A number of systems exist currently, incorporating varying degrees of sophistication. Broadly, these can be split into two groups:

- *Real-time systems* can "see" the position of all relevant trains on the network, can compare them to a timetable, if necessary, and can work out any changes in arrival/departure times or stopping patterns that may arise as a result of delays, etc. Examples of some real-time systems are shown in Figure 6.11.
- *Offline systems* simply display information from a fixed timetable. Generally, these types of systems require constant manual intervention to ensure that the information presented is relevant. The advantage of these systems is that they are generally cheap to build and maintain because they contain fewer system elements that can fail. The disadvantage is that a human is required to nurse the system constantly to keep it working.

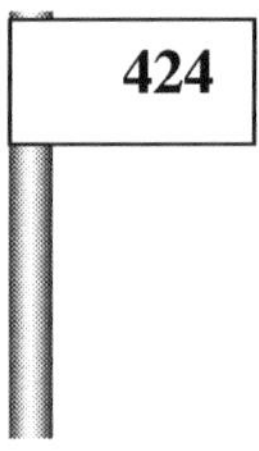

FIGURE 6.10 Location marker showing number of kilometers from a particular location.

FIGURE 6.11 (See color insert following page 154.) QR passenger information display system — an example of an online, real-time information system.

6.4 A WORD ABOUT SPAD

Railway signals are safety critical because their sole reason for existing is to ensure the separation of trains and therefore to prevent collisions between trains. However, because humans drive trains and humans make mistakes, it follows that signals are not always obeyed. One only has to conduct a newspaper search for the words "Ladbroke Grove" to come to the realization that, indeed, despite the best intentions of the signal design engineer, trains can and do collide because of missed signals.

The phenomenon of signals passed at danger (SPAD) is actually quite an ancient one; it is related to the (normally beneficial) intrinsic ability of the human to develop automated skills and to rely on such skills for navigating and maneuvering. Ordinarily, this mode of operation is useful because it allows the operator consciously to perform other cognitive functions instead of devoting all attention to driving the train. However, the problem of SPAD arises when visual cues are missed (such as the preceding yellow signal) or expected patterns of operation do not arise (for example, when a particular timetabled movement of a preceding train does not occur on time).

Invariably, very few SPAD are intentional events, but all are potential collisions and many railways are waking up to the realization that these human-factor events need to be prevented. Unfortunately, wayside signage will always deliver intermittent safety advice to drivers; i.e., a sign or signal is visible for only a certain amount of time as the driver approaches it. Once the train has passed the sign or signal, the information is then available only in the short-term memory of the driver, which can be expected to fade within a few seconds if it is not refreshed with some kind of reminder device or technique. However, even with the existence of driver reminder devices (e.g., AWS or the DRA), the device can often be ignored. There is enough cognitive load on the driver.

Although much industry research is being undertaken in the areas of human factors in the prevention of SPAD, as well as in the design of engineering solutions such as automatic train protection, the long-term solution lies in advanced signaling systems such as those described in the following section. However, the financial case for replacing the existing wayside signal infrastructure is rarely demonstrable using conventional accounting techniques, so railways are forced to work with the existing infrastructure while risk managers are forced to invest significant energy into developing risk management strategies to reduce SPAD.

Therefore, the state of the art in SPAD prevention in areas in which conventional wayside signaling exists (i.e., where there is no financial case for replacement) revolves around the following key components:

- SPAD awareness programs for train crews.
- Identifying and managing multi-SPADed signals and drivers.
- Improving design of conventional signals and operational processes to assist drivers in identifying and acting on signals at the appropriate times.
- Identifying environmental factors affecting situational awareness of drivers.
- Improving real-time and offline communication.

At the time of writing, the world's best practice in the reduction of SPAD in conventionally signaled territory (i.e., without the use of automatic train protection) is a 60% reduction in SPAD over a 5-year period. This type of reduction has been achieved in the U.K. (Railway Safety, 2003) and in Queensland (Borodin, 2003), using almost exclusively a soft-systems approach. This author considers that, for a conventionally signaled railway with no previous SPAD reduction program, an 80% reduction should be possible using a soft-systems approach over approximately a 10-year period (Borodin, 2003).

6.5 THE FUTURE OF RAILWAY SIGNAGE

By its very nature, railway signage is intermittent information provided from the trackside. Train drivers have access to that information for only a certain period of time while they travel toward it. Once they pass the sign or signal, they can no longer refer to it. If the information is safety critical, then the risk is that it can be forgotten and the train can proceed past its limit of authority, at which point no further guarantee of safe separation can be offered. Furthermore, wayside signage (particularly signaling) is extremely expensive to install and maintain, particularly in geographically large networks such as those extant in Australia. Wayside equipment is prone to damage from storms, floods, and vandalism.

Reasons such as these have prompted many railways to seek better ways of delivering safety-critical information to train drivers. At the time of writing, several systems for in-cab signaling have been designed and implemented, with varying degrees of success and applicability (Schmidt, 2002). Some examples include the TVM90 system in use on the French TGV lines, the proposed European ERTMS standard, and the German FFB and LZB systems. These systems have three basic features in common:

- They display target speed, actual speed, and distance to go.
- They intervene if the train's actual speed is outside the allowed target speed.
- They do not rely on wayside signaling.

One of the preceding systems, ERTMS, was proposed and designed to be interoperable with all existing European signaling systems (ERTMS, 2003). In Western Europe alone, there are 24 different systems for the automatic protection of trains, so interoperability is a primary concern of European rail operators. At this time, it is in an advanced state of testing and implementation at various locations throughout Europe. At its most advanced level, it is capable of providing continuous safe working information, including a moving map of the upcoming terrain, advanced GSM-R telecommunications, and remote intervention.

This in-cab information delivery paradigm is the future of railway signage and, indeed, it is the state of the art. Railways in continental Europe, such as Swiss Federal Railways and SNCF in France, have tested such systems on operational routes successfully (Barnard, 2002). These systems move beyond the need for wayside signage by bringing information to the operator's direct field of vision. Furthermore, they have the potential to make supervisory interventions in the case of human error or even to become complete autopilot systems that replace the function of the driver completely. Such in-cab information systems are discussed in the context of the road environment by Regan later in this book.

6.6 CONCLUSIONS

This chapter has attempted to explain something about the use of signage on railways. The unfortunate truth is that a single book could never successfully explain every signage treatment in the area of railway signage, even for a single railway. This is due to a constantly changing art of signage as new people enter each railway organization and bring with them new ideas, in addition to changing business requirements. If the author has at least been successful in stimulating the reader to dig deeper into this vast and curious topic, then it can be considered that this chapter has been successful in its aims.

FURTHER READING

IRSE Proceedings 2001–2002; Institute of Railway Signal Engineers.
http://www.signalbox.org/
http://www.trainweb.org/railwaytechnical/sigind.html
http://www.qr.com.au
http://www.railwaysafety.org.uk
http://www.networkrail.co.uk

REFERENCES

Barnard, R.E. (2002). IRSE first to visit French ERTMS test track 23–24 November 2001. *IRSE Proc.*, 2001–2002.

Borodin, A. (2003). Reducing train collision risk by 80% through pro-active management of SPAD. Lloyds List DCN. 3rd Annual Australian Rail Safety and Maintenance Conference Notes 2003: 20 February 2003, Sidney.

ERTMS (2003). ERTMS History at http://www.ertms.com/history.html, 22 December 2003.

Faith, N. (2000). *Derail: Why Trains Crash.* pp, 45–50. U.K.: Channel 4 Books.

Railway Safety (2003). Railway Safety and Standards Board UK; Safety performance report 2001/02; http://www.railwaysafety.org.uk; downloaded August, 2003, London.

Schmidt, F. (2002). Train control research in Europe. *IRSE Proc.*, 2001–2002.

7 Airport Signing: Movement Area Guidance Signs

Kirstie Carrick, Peter Pfister, Robert Potter, and Roy Ng

CONTENTS

7.1 Introduction: Why Are There Airport Signs, Anyway? 95
7.2 Standards for Airport Signing 98
  7.2.1 Summary of ICAO Annex 14 Recommendations 98
7.3 Human Factors Issues 100
  7.3.1 Location of Signs 101
  7.3.2 Size of Signs 101
  7.3.3 Technical Difficulties 102
  7.3.4 Environmental Influences 102
  7.3.5 Situational Awareness and Workload Issues 103
7.4 Runway Incursions 104
7.5 Interventions 105
7.6 Airport Signing: Case Study 108
7.7 Conclusions and Practical Implications 112
References 112

7.1 INTRODUCTION: WHY ARE THERE AIRPORT SIGNS, ANYWAY?

When the Wright brothers flew, they took off against the wind — whichever way it blew. This technique was followed for years, with grass airfields allowing aircraft to use any direction, as indicated by a windsock. Even as recently as the Second World War, airfields were grass and any direction could be used. As aircraft became larger and heavier and all-weather operations become more important, prepared smooth runways capable of carrying the loads imposed by large aircraft were required. As a consequence, paved taxiways were needed to get the aircraft from parking areas to the runways for takeoff and back after landing. Runways were obviously best aligned with the predominant wind direction.

0-415-31086-5/04/$0.00+$1.50

Growing complexity of the system led to the need for naming conventions for runways and taxiways. Runways are identified by compass direction; for example, an east–west runway is referred to as 09/27. Landing or taking off to the east is referred to as runway 09 (zero-nine), and if to the west, runway 27 (two-seven). Given that a pilot lands or takes off into the wind if at all possible, only one runway direction is used at a time. Taxiways leading to and from the runway are usually lettered and referred to by phonetic alphabet, thus taxiway A (alpha), B (bravo), etc., except that I and O are avoided because of possible confusion with the numbers 1 and 0, in written form.

To provide for different wind directions, an airport may have a second runway aligned in a different direction. Most often, these two runways will cross each other. With four runway ends to provide with taxiways, the map of the airport has become more complex. There might also be provision for an aircraft to enter or leave the runway at points between the two ends, so taxiways can intersect and cross the runways.

As air travel has increased, more traffic movements may have led to congestion on the airport. Perhaps the airport owner has had another runway constructed, parallel to the runway most often used but sufficiently far from it. Now 09L/27R and 09R/27L (for left and right when facing in the specified direction) may exist, as well as many more taxiways to feed aircraft to the runway and to get them from the runway to the terminal gates. Very large airports may have several terminals with 60 or 70 (or more) gates at each, multiple parallel runways in several directions, and hundreds of aircraft arriving and departing each day.

In these complexes of taxiways and runways, it is important to know which taxiway is which and, thus, where it goes. Airport diagrams give a plan view of the airport but signs are essential for the pilot to identify where the aircraft is and what intersection is being approached. Airport diagrams can become cluttered with information and have scaling problems; they are standardized at A5 page size, making them hard to interpret, especially for larger, more complex aerodromes (see Figure 7.1).

An analogy of a parking garage, to which most readers might more readily relate, may help explain the type of cognitive workload involved in taxiing a large aircraft in low visibility conditions at night from a runway to the terminal gate via a complex series of taxiways. Imagine someone driving a car in a large multilevel parking garage and seeking a parking space on a specific sign-posted level after receiving instructions by two-way radio from the parking attendant. Imagine also that, while driving, he is aware that other vehicles are traversing the garage, some moving up floors, some down. Add to this situation that it is night yet the car park has no lighting and also that thick fog is present so that, although the driver is using the headlight beam of his vehicle, he can see only a short distance ahead. Signs are at floor level and he may see them only relatively briefly. The co-driver is using a checklist to carry out various assigned functions related to the operation of the vehicle and the driver is required to be an active participant in some of these checklist actions. The parking attendant calls on the two-way radio to ask the driver's location. While endeavoring to ascertain the floor and ramp number, the driver is suddenly confronted by an oncoming vehicle coming down a ramp while his car is going up the same ramp. Fortunately, both vehicles stop before an impact occurs. No reverse gear is fitted to either vehicle, so the drivers sit there until tow trucks arrive to tow their vehicles away — assuming they can find them, of course.

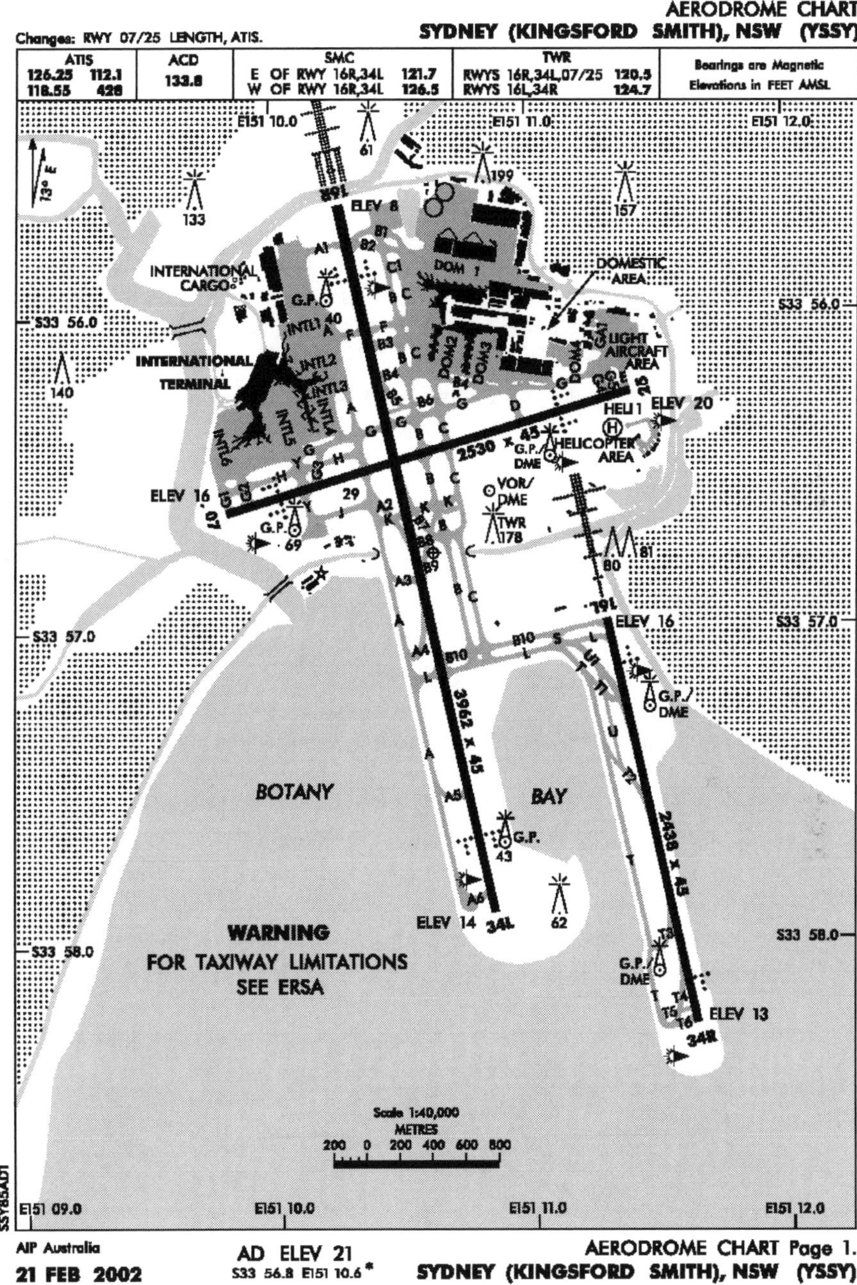

FIGURE 7.1 Aerodrome chart of Sydney Airport.

7.2 STANDARDS FOR AIRPORT SIGNING

The International Civil Aviation Organization (ICAO), part of the United Nations organization, monitors, advises on, and directs all matters regarding international civil aviation. It has determined a set of conventions with regard to the design, use, and positioning of signs on airports, known as movement area guidance signs, or MAGS.

7.2.1 Summary of ICAO Annex 14 Recommendations

The primary source of ICAO recommendations is Annex 14: Standards and Recommended Practices for Aerodromes (ICAO, 1999). A standard is defined as "any specification for physical characteristics, configuration, … the uniform application of which is recognized as *necessary* for the safety or regularity of international air navigation and to which Contracting States will conform under the Convention …" (ICAO, 1999, p. vii). A recommended practice is defined as "any specification for physical characteristics, configuration, … the uniform application of which is recognized as *desirable* for the safety, regularity or efficiency of international air navigation and to which Contracting States will endeavor to conform …" (ICAO, 1999, p. vii).

Airport operational signage forms a subgroup of visual aids for navigation within Annex 14 and Part 4, *Visual Aids*, of the ICAO Aerodrome Design Manual (ICAO, 1993). Two types of signs are described:

- *Mandatory instruction signs* identify a location where authorization from the control tower is required before proceeding. An aircraft (or ground vehicle) must stop and call the tower ground controller for permission to continue across the runway. The sign (see Figure 7.2), in white lettering on a red background, usually consists of the numbers of the runway being approached. Specific markings on the taxiway reinforce the point and indicate where the aircraft must stop. The signs and the markings should be co-located.

- DO NOT CROSS UNLESS CLEARANCE HAS BEEN RECEIVED.
- Located on taxiways at runway intersections and at runway/runway intersections

FIGURE 7.2 Runway hold position signs. (Courtesy of Lambert–St. Louis Airport, www.lambert-stlouis.com/slides/index.htm.)

• Location signs identify the taxiway on which the aircraft is located.

• Consists of yellow letters on a black background, bordered in yellow.

FIGURE 7.3 Taxiway location signs. (Courtesy of Lambert–St. Louis Airport, www.lambert-stlouis.com/slides/index.htm.)

• Provides direction to turn at next intersection to maneuver aircraft onto named taxiway.

• Consists of black letters on a yellow background and a black arrow, bordered in black.

FIGURE 7.4 Taxiway guidance signs. (Courtesy of Lambert–St. Louis Airport, www.lambert-stlouis.com/slides/index.htm.)

- *Information signs* identify a location or routing information. If referring to a specific location, the sign is the taxiway letter in yellow lettering on a black background, indicating that the vehicle is on this taxiway (see Figure 7.3). For other information, the letters indicating the crossing taxiways are in black lettering on a yellow background. If the sign is a combination, the location letter is usually in the middle with other taxiway letters on either side. Arrows may be included to indicate the angle of intersection (see Figure 7.4).

Because use of mandatory instruction signs requires the presence of a control tower, they may be limited in application for regional airports where air traffic control facilities are not present.

The use of movement area guidance signs is a standard and thus considered necessary at international airports, some of which may not fully comply with the standard while others may, depending on the country. In the U.S., the Federal Aviation Administration (FAA) uses advisory circulars to specify sign systems and standards for signs and for taxiway and runway lighting. The intent is that all certified airports have the same standard of signage and markings. The usage, colors, and placement of signs are the same as specified by ICAO. In other countries, the extent of compliance with ICAO recommendations varies; certainly, in Australia, it is common for smaller regional airports to have no signs.

7.3 HUMAN FACTORS ISSUES

Human factors associated with airport movement area guidance signs (MAGS) and their usefulness fall into three categories: (1) ergonomics of the signs and the aircraft; (2) capacity of the people using them to see, interpret, and use the signs correctly; and (3) organizational issues associated with airport activity.

Design and content of airport MAGS are specified in ICAO documentation and were discussed earlier. The design specifications are internally consistent and clear when the reasoning for them is understood. Ergonomic problems include the low level of the signs, which are easy to read if one is at 6 to 12 ft above ground. The cockpit of a 747 is 33 ft above ground level, which is equivalent to reading road traffic signs from the third floor of a building (see Figure 7.5). The signs are generally flat faced, not angled upwards which would be easier to read from a height but which would sometimes be subject to reflection and glare. The view from an aircraft cockpit is not particularly good in larger passenger aircraft because windshield size has been reduced to allow room for the flight deck instrumentation panel. In larger aircraft, it is also difficult to see things below the cockpit because of the forward shape of the fuselage. In wet weather on the ground, windshield wipers may be used; however, these do not wipe the entire windscreen area, so the viewing area is further reduced.

In addition, although the logic of the ICAO recommendations is clear once understood, it is not clear that pilots are ever trained in reading the signs or

FIGURE 7.5 View from B747 classic. (Courtesy of Boeing.)

instructed in their logic. Anecdotal evidence from pilot discussions (Krey, 2000) is that this is never taught and it is haphazard whether individual pilots comprehend the conventions.

ICAO-compliant signage appears at airports that conform to ICAO standards or recommendations. Not all airports are required to be compliant — only international airports. In fact, not all international airports use ICAO-compliant signage. A specific country may not be a signatory to the ICAO Convention, or it may have registered a "difference" to the requirements. As mentioned previously, regional and smaller local airports and airstrips may not have any signs or may have signs that are not designed as recommended by ICAO.

The purpose of MAGS is to ensure that pilots know where they are and thus can move from gate (or parking spot) to runway and from runway to gate in a safe and efficient manner, complying with the instructions of the ground controller. The first requirement is that the signs are legible to the pilot. Similar to many of the recommendations in other chapters of this book for signing in other transport domains, ergonomic considerations follow.

7.3.1 Location of Signs

Apart from the runway number and "piano keys" painted on the runway surface (the white stripes at each end that indicate the threshold of the runway), it is not recommended that MAGS be on or near a runway. This is to avoid unfortunate collisions at high speed if an aircraft should run off the runway when landing or taking off. Thus, generally once a pilot is on a runway, no signs indicate where he is. At some airports, runway signs and "distance to run" signs are adjacent to the runway, to indicate how far it is to the end of the runway; at some airports, taxiways may have an identification painted on the surface.

MAGS are located adjacent to taxiways, but the signs must be a specified distance from the edge of the taxiway (the distance is a compromise between visibility and clearance for aircraft) and located before an intersection, usually on the left side of the taxiway (the captain's side). However, it can become confusing in a sequence of intersecting taxiways to determine whether the sign relates to the intersection just passed or coming up. Mandatory signs and associated markings should be co-located, but when they are not, confusion can arise.

7.3.2 Size of Signs

ICAO criteria specify the size of signs to be used and their height above the ground. Clearance must be sufficient for the wing or engine nacelle of an aircraft to pass over the sign and the sign must be frangible, that is, must collapse if hit by an aircraft rather than damage the aircraft. ICAO criteria specify lighting uniformity and high illumination. These two criteria may not be easily achieved, particularly when the signs are large. The larger the sizes of signs, the more light output will be required to ensure that they are sufficiently conspicuous (i.e., more signage area will require more light in terms of illuminance/m^2).

Taxiway (and airfield) signs are lit by florescence or incandescent light source on a series circuit. Florescent light is normally in linear bands and incandescent light is from a point source. The design objective is to achieve a balance between the two criteria. Inadequate illumination or lack of lighting uniformity on the signs should be avoided because either deficiency can make the sign harder to read and thus have an impact on aircraft operational safety.

7.3.3 Technical Difficulties

Lighted taxiway signs are normally powered by a series circuit at 6.6 A in the airfield. A significant voltage drop makes it uneconomical for a normal 240-V supply circuit on what is often an extensive airfield network. Such a series circuit is not particularly suitable for florescent lighting, thus necessitating incorporation of an electronic ballast for the florescent light system. Although this type of light has an affinity for light uniformity, its application for airfield signs has limitations (see Figure 7.6).

7.3.4 Environmental Influences

Weather conditions can impede detection of signs in extreme cases. Heavy rain, fog, or snow can restrict the vision of pilots, especially if the flight deck is high above the ground. The development of aircraft automation is such that a modern airliner can land safely in conditions of very low visibility. The irony is that, once on the ground, it becomes very difficult to move the aircraft off the runway if low visibility is such that the pilots cannot see the ground. Runway and taxiway lighting (set into the surface) reduces the problem to some extent, but determining one's exact location requires the ability to see the location signs beside the taxiways.

FIGURE 7.6 (See color insert following page 154.) Taxiway signs on location. (Courtesy of Bo Wiberg, Swedflight Design Group: www.swedflight.com.)

In colder climates, these can be obscured by snow banks, even though the runway may be cleared.

Nightime taxiing has its own problems, as well, because visibility across the airport is more difficult. Runways and taxiways have inset edge lights and center line lights for assistance at night. These are yellow and white for runways and blue and green for taxiways.

7.3.5 Situational Awareness and Workload Issues

As seen in other places in this book concerning similar issues for road and rail signs, when an aircraft is moving, signs offer information (or instructions) to air crew for a limited time frame. If distracted during this "window of exposure," then the air crew may miss the information or not have sufficient time to assimilate and process the information adequately.

Workload may be high during taxi as preflight checklists are carried out; communication with the tower may be heavy as clearances are revised and sequences are determined. Postlanding, workload may be heavy as after-landing checks, locating the correct gate, and communication with the company or gate officials occur. During taxi, the pilot (or one of the pilots) must also find and read a sign to check position on the taxiway system, taking attention away from other tasks; if the other tasks hold the pilot's attention, relevant signage may be missed or misinterpreted. Some pilots prefer to use airport diagrams, but external features are still needed to check position. It is in this situation that runway incursions can occur. To add further confusion, airport diagrams are not always accurate. Anecdotal evidence suggests that some diagrams do not show airport expansions (Krey, 2000) because information can be slow getting to Jeppesen for update of the diagrams (Jeppesen supplies airport diagrams worldwide and provides updates regularly).

Distractions encountered by air crew during taxiing may include:

- Maintaining control of the aircraft, which may be difficult or at least a task requiring concentrated effort, particularly in strong wind conditions.
- Ensuring that the aircraft is on the taxiway because taxiways are relatively narrow for large aircraft and, in the case of four-engine aircraft, the outboard engines overhang the areas beyond the taxiway (including the signs).
- Communication within the flight deck.
- Communication with the cabin or persons entering the flight deck.
- Communication with the control tower or other aircraft.
- Communication with apron controllers or company representatives during maneuvering to the terminal in the apron area.
- Carrying out preflight checklists and preflight actions (associated with the checklist) when taxiing for takeoff.
- Carrying out after-landing actions and then the after-landing checklist during taxi to the terminal after landing.
- Maintaining track of current location and progress to the intended destination on the airport movement area (i.e., maintaining situational

awareness is a source of cognitive workload and thus a "distraction" when approaching signs).

- Unexpected approach of other aircraft or ground vehicles.
- Advised approach of aircraft or ground vehicles (which combines the distraction of communication with the tower with the acquisition of the aircraft or ground vehicle).
- Assessment by the flight crew of aircraft abnormalities prior to takeoff.

When situational awareness breaks down, the possibility for a conflict between two aircraft on taxiways or for a runway incursion is greater.

7.4 RUNWAY INCURSIONS

Normally, most regulators will defer to definitions and specifications issued by the ICAO. To date, however, ICAO has not issued a definition of runway incursions; thus, each national administration has determined for itself what constitutes a runway incursion. Australia and Europe take a simple view: Airservices Australia (the authority that administers airports and air traffic services) defines a runway incursion as an "Unauthorized Entry to an Active Runway Strip by an Aircraft, Person, Animal, Vehicle, or Equipment" (Airservices Australia, 2002). The European Joint Aviation Authority definition substitutes "unintended" for "unauthorized" (Eurocontrol, 2003).

The FAA in the U.S. distinguishes between runway incursion and surface incident, the latter relating to taxiways and tarmac areas other than the runway. Its definition also includes reference to the effect of the incursion on the safety of flight, creation of a collision hazard, or loss of separation. Furthermore, runway incursions are categorized as Category A, B, C, or D depending upon the level of safety threat resulting from the incursion. The U.S. National Transportation Safety Board (NTSB) has declared runway safety to be a primary target for improvement. The FAA has been running a campaign regarding runway incursions for several years, including the creation of an Office of Runway Safety within the FAA. Australia, Canada, and Europe have also taken up the issue.

Runway incursions increased steadily in the U.S. between 1993 (186 reported incursions) and the peak in 2000 (431 reported incursions). Since then, there has been a drop, with 383 in 2001 and 336 in 2002; however, these figures are still higher than any pre-2000 year (FAA, 2003a). Part of the increase shown may be due to an increase in reporting as much as any increase in events, because part of the FAA strategy regarding incursions is to encourage reporting of events. Of the categories given, pilot deviation is predominant — consistently between 56% and 60% of incursions. The other two categories are operational error and vehicle or pedestrian incursions, with the remaining incursions fairly evenly split among the categories. This categorization does not allow any assessment of the causes of the incursions, of course.

Elsewhere, Canada reported a 145% increase in runway incursions between 1996 (107 incursions) and 1999 (262) and noted that, in many locations where incursions were frequent, the local airport authority had taken steps to reduce the incursion rate, often related to deficiencies in signs, markings, and vehicle control (Transport Canada, 2000).

Following a runway incursion survey, the European authority reported that 48% of runway incursions were attributed to pilot deviation and a further 15% to joint pilot/ATC (air traffic control) contributions, with 12% operation errors, and the remaining 25% were due to vehicle or pedestrian deviations. They did note that the involvement of signage and markings was unclear from the survey data (Eurocontrol, 2002).

In Australia, figures for the 5 years to 2000 show an overall declining trend in incursions, from 111 in 1996, 95 in 1997, 121 in 1998, and 106 in 1999 to 101 in 2000. Of these, 71% were made by general aviation (GA) flights (excluding charter operation). The highest number of incursions was at Bankstown Airport, in western Sydney, frequented by GA operations (King, 2002). A survey by Airservices Australia (2002) asked pilots and air traffic controllers at Sydney (Kingsford-Smith) airport what caused runway incursions (many respondents gave multiple responses; thus, the total is more than 100%):

- Airport layout and complexity — 67%.
- Pilot experience — 61%.
- Airport signs — 56%.
- Runway and taxiway markings — 48%.
- Airport lighting — 45%.
- ATC procedures (including LTOP) — 41%.
- Language proficiency — 41%.
- A/G frequency congestion — 40%.

These Australian pilots thought that airport signs are a factor in runway incursions. This finding is also reflected in an online survey of pilots in 2000, in which *"inadequate and/or confusing airport signs and markings"* was indicated most frequently as the cause of runway incursions (Transport Canada, 2000). However, it may be that the problem is not so much the signs as the pilot's lack of understanding of signs and the conventions governing their positioning. The FAA claims that "airport marking, lighting, and signs have been standardized, at certificated airports with nearly virtual consistency throughout the nation" (FAA, Dec. 1999, p. 2).

The FAA reports that at towered airports, runway incursions tend to be due to pilots who take off from or enter a runway despite acknowledging instructions from the tower to the contrary. Other causes are air traffic controllers allowing aircraft to get too close together or drivers or pedestrians on the tarmac who do not communicate with the tower or apparently ignore instructions from the tower (Rodgers, 2002). These conclusions do not explain why such loss of situational awareness occurs, nor are signs explicitly mentioned.

7.5 INTERVENTIONS

Many forms of technology-oriented and nontechnological intervention are promoted. Because most incursions involve GA aircraft (72%; FAA, 1998), the intervention program in the U.S. has focused a great deal of attention on GA pilots, including educational kits, brochures, and posters. Most of this material emphasizes the following advice (FAA, 2000):

- Listen on the radio to what is going on.
- Transmit clearly.
- Write down taxi clearances.
- Keep an airport diagram or map handy.
- Admit if you are lost.
- Keep a sterile cockpit (ask passengers to keep quiet in small planes).
- Understand airport signs and markings.
- Never assume that the clearance means nothing is coming; continue to look along the runway before crossing or entering.
- Stick to safe procedures.

Similar advice has been directed at GA pilots in Australia (King, 2002). Clearly, it is recognized that pilots need to be aware of the meanings of airport signs and markings as well as maintain situational awareness regarding operations at the airport.

The JAA Strategic Safety Initiative in Europe appears to be a coordinated approach, with a European Action Plan for the Prevention of Runway Incursions. The plan addresses ATC phraseology and procedures, pilot procedures, driver training and procedures on airports, implementation of ICAO Annex 14 (signs and markings), and development of individual runway safety programs in various states. Evidently, the concern is that not all airports meet the ICAO standards for airport signs and markings and the commitment is to move toward raising all airports to that standard.

The U.S. has issued a comprehensive document addressing runway safety, the Runway Safety Blueprint 2002–2004. Although the overall impression is of emphasis on technological intervention in ATC and the flight deck, it is also apparent that attention is being given to education of GA pilots and airport vehicle drivers and development of runway safety programs (FAA, 2003c). It is also reported that the FAA has sponsored research into improving the visibility of runway holding position markings because these markings seem to be ignored during runway incursions. It is suggested that changes to the colors used, extending the lines beyond the taxiway onto the "shoulders", and adding signs with the markings (as recommended by ICAO) would help. The FAA is moving toward validating these recommendations (FAA, 2003b).

More technological interventions that have been developed, trialed, and, in some cases, introduced include various aircraft location systems: the U.S. FAA AMASS (airport movement area safety system) and ASDE-X (airport surface detection equipment — model X), multistatic dependent surveillance (MDS), and runway warning alert system, as well as laser lights at hold points and displays for cockpits incorporating GPS (global positioning system) and ATC datalink information.

AMASS enhances radar surveillance of aircraft and airport surface vehicles at high-activity airports. It provides air traffic controllers with automated alerts and warnings of potential runway incursions and other hazards, thus improving safety (FAA, 2002a). Currently, 29 U.S. airports have AMASS operational equipment (FAA, 2003b). Because it uses radar images, AMASS is dependent upon vehicles and aircraft enabling their transponders for identification but can still locate (but not identify) vehicles without transponders. The system has problems because the radar system is "line of sight" and it is difficult to separate aircraft in the radar shadow of others or in terminal and hangar areas. The FAA has reduced its use to runways only (not taxiways), but the concern is that if alerts are issued only after a runway incursion has occurred, time to warn any other aircraft involved will be short (IASI, 2003).

The ASDE-X systems are for use at airports without AMASS capability, although some airports will have both. Currently seven airports have the system; the intent is to install at 25 airports. The system indicates aircraft position and identification information overlaid on a color map showing the surface movement area and arrival corridors for the particular airport (FAA, 2002a).

A number of enhancements are being developed by various engineering organizations to address this time lag and to evaluate the effectiveness of the system, under the auspices of NASA (for example, Cassell et al., 2000). It is apparent that the NTSB also has had concerns regarding the ability of the AMASS system to give sufficiently timely warnings (NTSB, 2000). The FAA has also sponsored evaluation of human–system issues related to ASDE-X and similar systems (e.g., Parasuraman et al., 2002).

The multistatic dependent surveillance (MDS) system locates vehicles equipped with transponders by using at least three nonrotating sensors, located at different positions on the airport, that resolve an aircraft's position from its transponder signal. More sensors might be needed, depending on the obstructions on the field and extent and accuracy of coverage required (Sensis, 2003). MDS systems have been installed at a number of airports in the U.S., Canada, and Europe, including Dallas–Fort Worth and Heathrow. This system seems more robust and accurate than the AMASS and ADSE-X systems and can be utilized for aircraft separation over varying areas and in areas not easily managed by normal radar. However, it depends upon transponder signals from aircraft or ground vehicles and is blind to nontransponder-equipped vehicles (or those in which the transponder is not activated).

Another proposal for ground movement monitoring and control is the use of in-ground inductance loops called the runway warning alert system. This is the same sort of technology used to manage traffic lights; the presence of a large metal object over the loop embedded in the taxiway (in the airport scenario) creates a signal that indicates vehicle position to the ATC tower or other ground control facility (Asti, 2001). It is not clear whether this system would be able to identify individual vehicles or track them between the inductance loops. Its promoters note that it is less expensive than many of the radar-based systems, but it does not seem to be as flexible or informative.

The systems described are more concerned with providing information to ATC (or ground control) about aircraft movements so that they can issue appropriate instructions to avoid aircraft or ground vehicle conflict. It does not appear that any information is provided directly to the aircraft. Thus, it is unlikely that any of these interventions will replace airport signs because the aircraft (and ground vehicles) will still need to know where they are.

Additional technological interventions address the information needs of the pilot. One such proposal that would provide positioning information to taxiing aircraft utilizes GPS technology in the cockpit. It also requires retrofitting a large LCD screen to replace the standard flight instruments. In flight, the screen shows the standard displays on the top half of the 15-in. display and weather radar or TCAS (traffic alert collision avoidance system) information on the bottom half of the screen. When the aircraft lands, this part of the screen can be changed to show the airport diagram from the Jeppesen database and the aircraft's current position from a GPS receiver (Jensen, 2001). This is a development aimed at aircraft that do not currently have electronic flight instrument displays and is said to be affordable (less than U.S. $100,000). The predicted release date was March 2003, but as yet nothing further has been heard about the technology.

Research has also been reported on the use of a head-up display in the cockpit to show the cleared route across the airport. This would require input from the tower, perhaps via datalink technology that allows direct data transfer from tower to aircraft, in conjunction with some form of vehicle location system, such as ADSE or MDS (Young and Jones, 2001).

A more practical problem raised earlier in this chapter is the proposal to enhance existing runway markings in poor visibility conditions. Recently trialed in Alaska, a pair of laser projectors point at each other across the taxiways about 2 ft above the hold lines. In poor visibility the laser beams will be reflected from the fog or ice particles to produce an apparent barrier across the taxiway (Pemberton, 2002; FAA, 2002b). This sounds quite effective, at least for smaller aircraft; 2 ft above the ground is still a long way down from the cockpit of a larger Boeing or Airbus passenger-carrying jet. The evaluation of this system was carried out in November 2002 but as yet no publication of the evaluation has appeared.

7.6 AIRPORT SIGNING: CASE STUDY

Late in the evening of October 31, 2000, heavy rain and strong winds stirred up by typhoon Xiangsane swept across Chang Kai-shek Airport, near Taipei on the island of Taiwan. For the pilots of Flight SQ006, a Singapore Airlines B747 bound for Los Angeles, the view from the cockpit was difficult but not unusual for a wet, windy, dark night. The captain taxied onto a runway that he thought was the correct one, 05L, but that was, in fact, 05R. The take-off roll commenced and about 20 seconds later the aircraft hit concrete barriers and construction equipment, broke up, and caught fire; 83 people died (Aviation Safety Network, 2002; Singapore MOT, 2002a). A number of factors led to positioning of the aircraft on the wrong runway, among which are issues to do with the provision of airport signs and markings, especially the lack of any indication on the ground that the runway was closed and being used only for taxiing (RAES, 2002).

The captain taxied the aircraft toward the assigned runway, 05L, by following the green taxi lights set into the center of the taxiway, having briefed the taxi route with the crew with reference to the Jeppesen airport diagram. The diagram showed

05L and 05R as active open runways; however, the crew had access to NOTAMs (notices to airmen) that indicated part of 05R would be closed and that the remainder of the runway would be redesignated as a taxiway and runway markings would be changed. A more recent NOTAM (October 23, 2000; cited by Singapore MOT, 2002a) stated that the redesignation had been postponed. There was no clear statement about the current state of the runway markings at the time of the accident.

Taxiing toward the runway, the aircraft turned left from the west cross taxiway onto taxiway NP, which ran parallel to runway 05, at a point after taxiway NP had passed the construction area, off to the right of the intersection. The take-off clearance was given before the aircraft left taxiway NP, so it did not need to locate and stop at the hold point for the runway. The captain followed the lights from taxiway NP around to the right onto taxiway N1. Very shortly thereafter, he saw the lights curve to the right onto a runway past the threshold "piano keys." It never crossed his mind that he was entering 05R and not 05L. The runway sign on taxiway N1 indicating the next crossing runway as 05R was positioned on the left, before the runway as required, but not co-located with the 05R hold point, which was on taxiway NP. However, because the sign was also just after the intersection of NP and N1, it was difficult to see from an aircraft approaching from NP when all visual cues were leading to the right (Singapore MOT, 2002a).

The runway sign for 05L was on the left of the taxiway and much farther down taxiway N1 — well beyond 05R and, in fact, positioned beyond the hold point for runway 05L rather than co-located with it. The hold point for 05L was not lit as recommended by ICAO, with red lights set into the surface and yellow lights beside the taxiway. Thus, it was not conspicuous, especially from as far back as 05R. The green taxi lights and yellow center marking led onto runway 05R. A set of green taxi lights and a yellow center line should have led across the 05R piano keys toward 05L. In fact, four lights (16 are recommended for the distance involved) led to the 05L hold point. The first, just before the piano keys, was inoperative; the second, in the middle of the keys, was dim (possibly invisible in the rain); and the third was on the far side of 05R, more than 100 m from the point at which the lights began to curve to the right onto 05R (Singapore MOT, 2002a).

The large painted "R" on the runway just before the "05" was apparently not noticed as the aircraft moved over it. The first officer's comment that the PVD (paravisual display) had not lined up was dismissed by the captain with "Never mind, we can see the runway." The PVD indicates that the aircraft is lined up with the ILS localizer for the runway and is intended to help the pilot track the runway center line in low visibility conditions. It is not intended as a runway identifier and is not needed if visibility is sufficient to see the center line lights. Other in-cockpit displays, the primary flight display and navigation display, were not, and are not normally, used to confirm runway position, nor were any discrepancies particularly obvious (Singapore MOT, 2002a; also see Figure 7.7).

The analysis of the accident by the Singapore Ministry of Transport team (2002a) concluded that the runway edge lights for runway 05R were illuminated at the time of the SQ006 accident. The lights were white; taxiway edge lighting is blue. The center line lights were green, appropriate for a taxiway, rather than the white of an active runway, but had led from N1 onto 05R. As the MOT report states, "These

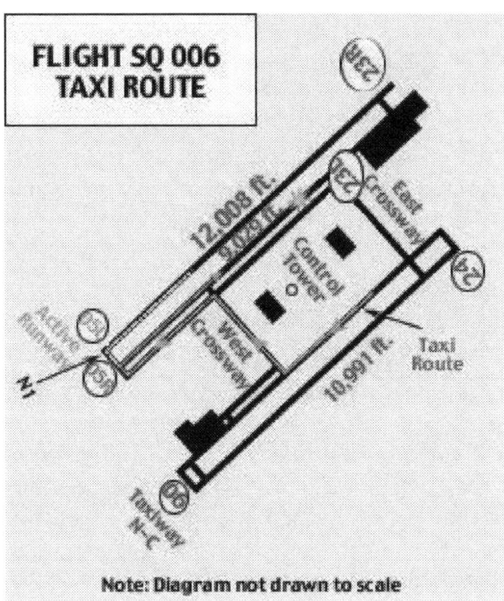

FIGURE 7.7 Taxi route taken by SQ006. (Courtesy of Air Traffic Cafe) www.airtraffic-cafe.com.)

taxiway centerlines provided a compelling visual pathway. There was no other continuous line of taxiway centerline lights visible to the crew which would have indicated that there was an alternative taxi path available to them" (Singapore MOT, 2002a, p. 17). Once lined up on the runway, the crew was presented with the cues that they expected of an operational runway: edge lighting and center lights. The lights on the construction equipment and the barrier just ahead of it were 1240 m down the runway and not visible in the rainy conditions.

Because runway 05R was inoperative and no plans to reopen it were apparent, the threshold markings (piano keys) and the runway number should have been painted out and the green taxi lights should have continued along N1 more conspicuously. In addition, no signs or markings on 05R indicated that it was closed, nor were any barriers across the runway. Such barriers are used elsewhere and could have been established on 05R. One such example is a towable barrier illuminated with red lights in the form of a line of lights across the runway and a cross vertically in the center, used at Charles de Gaulle Airport in Paris (Singapore MOT, 2002a). Significantly, some of these steps (removing the threshold markings and painting a yellow taxiway center line toward 05L, adding taxi lights, redesignating the runway as taxiway NC, and disconnecting the runway edge lights) were taken by the airport authority immediately following the accident (Singapore MOT, 2002b).

It appears that a number of cues led the pilots to believe that they were on runway 05L and there were no explicit cues to tell them that they were on 05R, a closed runway. They expected to see a runway, they were on a runway, and they saw the correct runway (see Figure 7.8 and Figure 7.9).

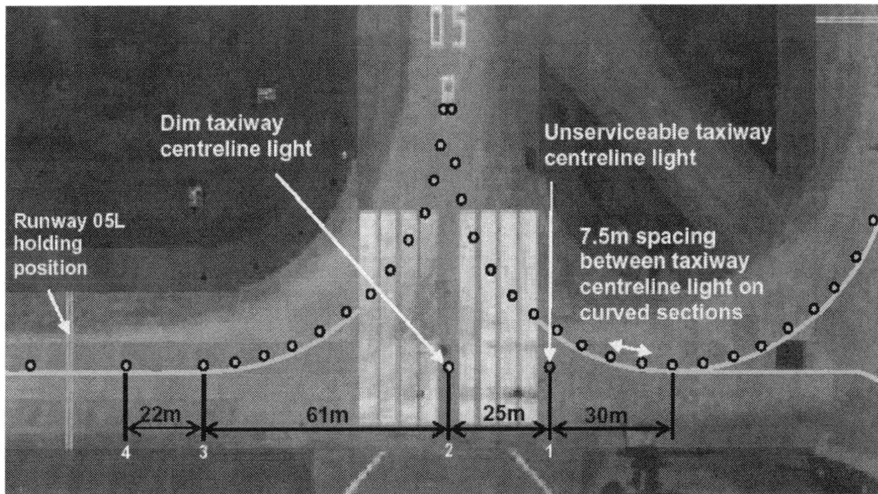

FIGURE 7.8 Runway lights and markings at 05R at the time of the SQ006 accident. (Courtesy of Singapore Ministry of Transport.)

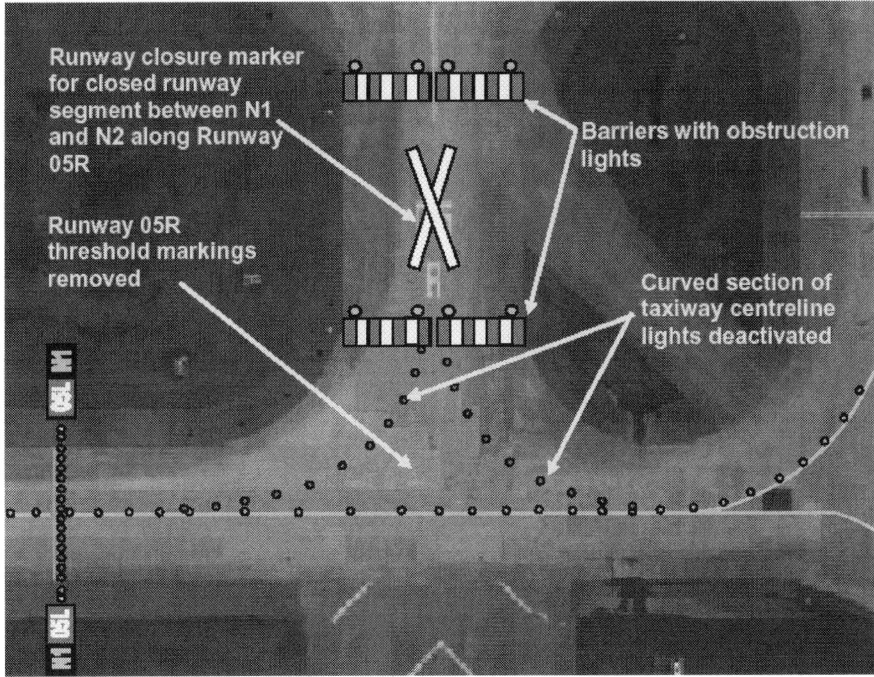

FIGURE 7.9 Runway lights, markings, and barriers showing 05R as closed. (Courtesy of Singapore Ministry of Transport.)

7.7 CONCLUSIONS AND PRACTICAL IMPLICATIONS

The intervention programs in the U.S., Canada, and Europe are broad-based approaches to dealing with the runway incursion problem; they address all users of airports and include attention to the provision of airport signage in accordance with ICAO recommendations. Australia is setting up a runway incursion task force within Airservices Australia to monitor and make recommendations with regard to runway incursions. The provision of signs at smaller regional airports in Australia is an issue for the individual airport owners and Airservices Australia, the relevant government authority. Other interventions, such as education and awareness, are being addressed through industry publications like *Flight Safety Australia.*

The more high-tech interventions discussed earlier also appear to be high cost and may be out of reach of small operators and small airport owners. Many depend upon the airport's having an air traffic control tower or some other form of surface movement control; therefore, this type of intervention is going to be established only at the larger, more traffic-dense airports where incursions are more likely.

The main problem remains one of human factors; interventions may be present and working, but the context of aviation activity at the airport remains the same. Situational awareness and cognitive workload will still push individual pilots (and air traffic controllers) to the limits of their capacity. The issue seems to be not so much with the provision of signs and markings as with getting pilots and drivers of ground vehicles to look for and use the signs, follow instructions, and maintain situational awareness under high workloads.

REFERENCES

Airservices Australia (2002). Runway incursion survey. Retrieved June 17, 2003 from http://www.airservicesaustralia.com/pilotcentre/Runwaysafety/Runway%20Incursion%20Survey%20Results%20Brochure.pdf.

Asti (2001). A viable solution to runway incursions. Retrieved August 21, 2003 from http://www.asticorp.com/rwas.htm.

Aviation Safety Network (2002). SQ006 accident description. Retrieved August 21, 2003 from http://aviation-safety.net/database/2000/001031-0.htm.

Cassell, R., Evers, C., and Yang, Y.E. (2000). Pathprox — a runway incursion alerting system. IEEE. AIAA 19th Annual Digital Avionics Systems Conference. Retrieved August 21, 2003 from http://avsp.larc.nasa.gov/pdfs/crp-fd-np14.pdf.

Eurocontrol (2002). Presentation on runway incursion survey results. Retrieved August 21, 2003 from pre5thRSSdata_020214.ppt at http://www.eurocontrol.int/eatmp/events/runwaypgme.html.

Eurocontrol (2003). The European action plan for the prevention of runway incursions. Retrieved June 17, 2003 from http://www.eurocontrol.be/eatm/agas/runwayincursions/actionplan.html.

FAA (1998). Runway incursion recommendations, RE & D subcommittee on runway incursions, FAA Aviation Research Organization. Retrieved June 17, 2003 from http://www1.faa.gov/aar/red/rep-min/runway_incursion.htm.

FAA (Dec, 1999). How it all began... *FAA Airport Saf. Newslett.*, 1(2), 1–3. Retrieved June 17, 2003 from http://www.faa.gov/arp/pdf/nwltr12.pdf.

FAA (2000). Ten ways to help prevent runway incursions. FAA poster presentation. Retrieved June 17, 2003 from http://dot.state.oh.us/aviation/FAA/RwyIncursions.pdf.

FAA (2002a). Runway safety blueprint 2002–2004. Retrieved August 20, 2003 from http://www1.faa.gov/nasarchitecture/blueprnt/2002Update/06-separation-systems2.htm.

FAA (2002b). Airport laser holding position enhancement project plan. Alaskan Region Runway Safety Program Office. Retrieved August 20, 2003 from http://www.alaska.faa.gov/ runwaysafety/Templates/Laser%20Plan.pdf.

FAA (2003a). Runway incursions by category, 1988 to 2002. Runway Safety Program Office. Retrieved August 20, 2003 from http://faarsp.org/xricats88-01.htm.

FAA (2003b). FAA runway safety report: runway incursion trends at towered airports in the United States FY1999–FY2002. FAA Office of Runway Safety. Retrieved August 20, 2003 from http://www faarsp.org/pdf/report3.pdf.

FAA (2003c). FAA Office of Runway Safety Web site. Retrieved August 20, 2003 from http://www.faarsp.org/index.shtml.

IASI (2003). A mess called AMASS. Retrieved August 20, 2003 from http://www.iasa.com.au/folders/a_mess_called_amass.html.

ICAO (1993). *ICAO Aerodrome Design Manual Part 4 – Visual Aids* (Doc 9157-AN/901 Part 4), 3rd ed. Montreal, Canada: ICAO.

ICAO (1999). *Annex 14: Standards and Recommended Practices for Aerodromes,* 3rd ed. Montreal, Canada: ICAO.

Jensen, D. (2001, Nov.). LCD: bigger than ever. *Aviation Today.* Retrieved August 20, 2003 from http://www.aviationtoday.com/reports/avionics/previous/1101/1101iss.htm.

King, R. (Jan.–Feb., 2002). Runway incursions, reducing the risk. *Flight Saf. Austr.*, 24–29.

Krey, N. (2000). Neil Krey's CRM developers forum. Retrieved January 12, 2003 from http://groups.yahoo.com/group/crm-devel/.

NTSB (2000). Press release: public meeting June 13, 2000. Retrieved August 20, 2003 from http://www.ntsb.gov/pressrel/2000/000613.htm.

Parasuraman, R., Hansman, J., and Bussolari, S. (2002). Framework for evaluation of human system issues with ASDE-X and related surface-safety systems. Retrieved August 20, 2003 from http://psychology.cua.edu/csl/recent-papers.cfm.

Pemberton, M. (2002). Laser lights may help prevent collisions on runways. Retrieved August 20, 2003 from http://www.redding.com/news/business/past/20021118 bus026.shtml.

RAES (2002). The crash of flight SQ006: pilot error or system failure? Lecture by Rob Lee for the Canberra Branch of the RAES, July 9, 2002. Retrieved August 20, 2003 from http://www.raes.org.au/Canberra/Previous/Jul-02.htm.

Rodgers, M.D. (2002). Research report — runway incursions. *HF Newslett.*, 02-14, 1–4. Retrieved June 17, 2003 from http://www.hf.faa.gov.

Sensis (2003). Multistatic dependent surveillance. Retrieved August 20, 2003 from http://www.sensis.com/docs/49/.

Singapore MOT (2002a). Analysis of the accident to Singapore Airlines flight SQ006, Boeing 747-412, 9V-SPK at Chiang Kai-shek Airport, Taipei, Taiwan, on 31 October 2000. Ministry of Transport, Singapore.

Singapore MOT (2002b). Press release — 26th April 2002: Singapore Ministry of Transport comments to the final report of the investigation into the SQ006 accident. Retrieved August 20, 2003 from http://www.gov.sg/mcit/newsroom_nav/p_land.htm.

Transport Canada (2000). Final report of the subcommittee on runway incursions. Retrieved August 20, 2003 from http://dsp-psd.communication.gc.ca/Collection/T52-104-2001E.pdf.

Young, S.D. and Jones, D.R. (2001). Runway incursion prevention: a technology solution. Presented at the Joint Meet. Flight Saf. Found. 54th Ann. Int. Air Saf. Semin.; *Int. Fed. Airworthiness 31st Int. Conf.; Int. Air Transpt. Assoc.*, November 5–8, 2001, Athens, Greece. Retrieved August 20, 2003 from http://avsp.larc.nasa.gov/pdfs/csrp14.pdf.

8 The Aging Eye and Transport Signs

Donald Kline and Robert Dewar

CONTENTS

8.1 Visual Problems of Aging Drivers 116
8.2 Age-Related Visual Decline 116
 8.2.1 Age-Related Refractive Problems 117
 8.2.2 Light, Glare, and Color 117
 8.2.3 Spatial Vision and the Aging Driver 118
  8.2.3.1 Acuity 118
  8.2.3.2 Static Contrast Sensitivity 118
  8.2.3.3 Dynamic Vision 118
 8.2.4 The Useful Field of Vision 119
  8.2.4.1 Useful Field of Acuity and Contrast Sensitivity 119
  8.2.4.2 Useful Field of View (UFOV) 119
 8.2.5 Eye Movements and Visual Search 119
  8.2.5.1 Aging Effects on Eye Movements 119
  8.2.5.2 Visual Search 120
8.3 Transport Signs and the Aging Operator 120
 8.3.1 Conspicuity 121
 8.3.2 Legibility 121
  8.3.2.1 Legibility: Symbol vs. Word Signs 123
 8.3.3 Glance Legibility 124
 8.3.4 Comprehension 125
 8.3.5 Perception–Response Time 125
8.4 Optimizing Sign Legibility 126
 8.4.1 Word Signs 126
 8.4.2 Symbol Signs 126
8.5 Strategies for Enhancing Signage for Older Drivers 128
 8.5.1 Strategies for Older Operators 128
 8.5.2 Strategies for Traffic Professionals 128
8.6 Conclusions and Future Research 129
References 130

0-415-31086-5/04/$0.00+$1.50

8.1 VISUAL PROBLEMS OF AGING DRIVERS

Although older drivers do not contribute unduly to traffic accidents overall, accident rates in terms of distance traveled increase markedly in the later years (Stamatiadis and Deacon, 1995; Ryan et al., 1998). Relative to their young and middle-aged counterparts, older drivers (e.g., those 65 and older) tend to be overrepresented in accidents involving turning or changing lanes, improper failure to yield the right of way, and failure to heed signs or signals (Huston and Janke, 1986; McGwin and Brown, 1999).

The elevated accident experience of older drivers appears to result from age-related changes on a wide range of interacting sensory, perceptual, cognitive, psychomotor, and health variables (Marottoli et al., 1998; McKnight and McKnight, 1999; Lyman et al., 2001). Given the critical role of vision in many driving tasks, it is not surprising that visual decline figures prominently in the challenges faced by older drivers (Shinar and Schieber, 1991) or that they show considerable insight into their visual problems. Drivers over age 55 surveyed by Yee (1985) reported a variety of visual problems, including decreased visibility in low illumination and difficulty seeing highway signs and markers. When Kline et al. (1992) surveyed drivers aged 22 to 92 years regarding their visual problems, several problems were found to increase with age: dealing with unexpected vehicles, judging and dealing with vehicle speed, reading dim in-vehicle displays, dealing with windshield problems, and difficulty reading signs.

Schieber et al. (1992) found that older drivers' visual problems were associated with objectively measured losses on contrast sensitivity. Kosnik et al. (1990) observed that older persons with self-reported difficulties on everyday nondriving visual tasks were also more likely than their cohorts of the same age to cease driving. Klein et al. (1999) found that performance-based and self-reported measures of visual function were consistently related to age and that self-reported visual functions were significantly correlated with objective visual measures. Approximately 39% of older drivers reported that visual decline had caused them to limit night driving. Burns (1999) surveyed British drivers aged 21 to 85 years to determine how navigation problems affected their mobility. The frequency of self-reported difficulties in finding one's way increased with age, as did the avoidance of unfamiliar places and routes. He concluded that the mobility of older drivers could be enhanced by the use of navigational aids such as maps, in-vehicle guidance systems, and better road signs.

8.2 AGE-RELATED VISUAL DECLINE

Sign detection and legibility are affected adversely by normal age-related visual decline (Kline and Scialfa, 1997) as well as by eye diseases prevalent among the elderly (Ivers et al., 2000; Brabyn et al., 2001). Legal blindness (best corrected acuity of 20/200 or worse in the better eye and/or a visual field smaller than 20 degrees) increases markedly in the later years. In the U.S., about 1% of those aged 70 to 74 are blind, as are about 2.4% of those 85 and older (Desai et al., 2001). The elevated prevalence of blindness is accounted for largely by four progressive disorders: age-related macular degeneration (ARMD), cataracts, diabetic retinopathy, and glaucoma (Desai et al., 2001). Visual impairments of more moderate degree, however, are more relevant for driving and transport operations because they are more prevalent and drivers with them are more likely to meet vision standards for licensure or certification. Drivers with mild visual impairments are also less likely than those with severe impairment to cease driving voluntarily due to reduced vision (Kosnik et al., 1990; Holland and Rabbitt, 1992).

8.2.1 Age-Related Refractive Problems

Refractive errors necessitating the use of corrective lenses are increasingly common among older observers. *Hyperopia* — the inability to focus on close displays — appears to increase in prevalence with age until about 70. Thereafter, the prevalence of *myopia*, the inability to focus on far stimuli such as signs (Lee et al., 1999), may increase. Astigmatism, a problem of irregular focus, appears to change in prevalence and orientation (i.e., axis) with age. Gudmundsdottir et al. (2000) found that the prevalence of astigmatism requiring ±0.75 diopters (D) or more of lens correction on the right eye increased from about 37% of those aged 50 to 54 to almost 90% in those over 80. Against-the-rule (clearest vision within ±15° of the horizontal axis) and oblique astigmatisms were increasingly common, while the prevalence of with-the-rule astigmatism (±15° of the vertical axis) declined. Older observers can avoid the associated loss of legibility for vertically and obliquely oriented sign elements by updating their optical prescriptions regularly.

By age 60, virtually all of the ability to increase the sphericity of the lens to focus near stimuli (*accommodation*) is lost (Bruckner, 1967; Kalsi et al., 2001). This normal recession of the near point of vision (*presbyopia*) is often first noticeable at about 40 years of age. Although the loss of near focus is correctable by spectacles with added spherical power (e.g., reading glasses or bifocals), displays at intermediate distances (1 to 6 m) may be less than optimally focused for older observers. Although trifocals and progressive lenses can provide more continuous distance correction, they also demand fairly precise alignment of the observer's gaze.

8.2.2 Light, Glare, and Color

Changes in the optic media are estimated to reduce retinal illumination at age 60 to about one third of its level at 20 (Elliott et al., 1990). As a result, legibility and object detection tasks are disadvantaged by low illumination among older observers (Burns et al., 1999). Brabyn et al. (2001) found that even older observers with good acuity might be impaired on everyday tasks involving low and changing light levels, glare, or low contrast.

Although increased light levels can enhance target legibility, older observers' greater susceptibility to glare makes it important to assure that light is appropriately directed. Olson and Sivak (1984) found that transient glare produced by headlamp reflection in the rear-view mirror, even at low-beam levels, affected forward visibility more adversely and for longer durations for old than young drivers. Elliott and Whitaker (1991) found the older eye to be slower also to recover good acuity after exposure to strong glare. These findings suggest that evaluation of the size and contrast of sign elements should include an assessment of the degree to which they counter older operators' susceptibility to and protracted recovery after glare exposure.

Selective absorption of short wavelengths by the senescent eye tends to degrade color discrimination for the blue–green region of the spectrum. Knoblauch et al. (1987) axis found that the decline in color discrimination along the tritan (i.e., blue/yellow) was exacerbated under reduced illumination. This deficit is generally greater for desaturated colors (Cooper et al., 1991). Because transport signs do not normally require discrimination of similar colors, age-related effects on their legibility are likely to be small except perhaps in cases of a visual disorder (e.g., ARMD or cataract) or for badly weathered signs.

8.2.3 Spatial Vision and the Aging Driver

8.2.3.1 Acuity

Static acuity, the smallest detail that can be resolved in stationary targets, declines after age 50 or so, even among healthy, well-corrected observers (Elliott et al., 1995). Among less select "epidemiological" populations, age-related acuity declines are even more robust (Haegerstrom–Portnoy et al., 1999; Klein et al., 2001). Attebo et al. (1996) found that best-corrected far acuity declined with age from the youngest age tested (49 years) onward. Older observers' acuity deficits are exaggerated by low contrast or luminance (Haegerstrom-Portnoy et al., 1999). Sturr et al. (1990) found that the acuity of 77% of drivers underage 65 was 20/40 or better at low luminance (2.4 cd/m^2). For those aged 65 to 75 and over 75, the corresponding proportions were 28 and 4%, respectively. Well-illuminated, high-contrast signs can go a long way toward minimizing the effects of older observers' acuity limitations on sign legibility.

8.2.3.2 Static Contrast Sensitivity

Age-related declines on the contrast sensitivity function (CSF), a measure of the observer's ability to discriminate light/dark differences for gratings, are most notable for gratings of intermediate and high spatial frequency (Elliott et al., 1990; Kline et al., 1983). This deficit is reduced but not eliminated by optimal lens correction (Owsley et al., 1983; Scialfa et al., 1991) or by elevated luminance (Owsley et al., 1983). Because real-world targets vary widely in size and contrast, CS measures are generally more suitable than acuity for predicting their detection or identification. For example, CS measures have been shown to better predict the detection and identification of common objects (Owsley and Sloane, 1987), self-reported visual problems of aging drivers (Schieber et al., 1992), and discriminability of highway signs (Evans and Ginsburg, 1985).

8.2.3.3 Dynamic Vision

Whether measured in terms of the smallest detail that can be resolved in a moving target (i.e., *dynamic visual acuity*) or in the contrast needed to discriminate a moving grating (*dynamic contrast sensitivity*), dynamic modulation exacerbates the legibility problems of older observers (Burg, 1966; Reading, 1972). The age-related decline in retinal illumination (Long and Crambert, 1990), as well as inadequate smooth pursuit gain and/or protracted temporal summation in the senescent visual system (Scialfa et al., 1992), may contribute to this deficit. Regardless of its origin, the decline in dynamic vision has negative implications for the ability of older operators of moving vehicles to extract information from signs.

8.2.4 The Useful Field of Vision

8.2.4.1 Useful Field of Acuity and Contrast Sensitivity

The normal falloff in acuity from the center to the periphery of the visual field is more pronounced among older observers. When Collins et al. (1989) compared the peripheral acuity of young and old observers with good central acuity (20/20 or better), the young were able to identify a 2.4-miniarc (20/48) target at eccentricities up to 30.8°, whereas older observers were unable to do so beyond 22.8°. The deficit was smaller when a larger (4.8-miniarc-20/96) target was tested. An age-related decrement has also been shown for peripheral contrast sensitivity (Crassini et al., 1988). Such reductions in the useful field of spatial vision may impair older operators' ability to notice or process peripherally presented sign information. It has been shown, at least, that the traffic accident and conviction rates of drivers with binocular visual field losses are about double those of observers with normal fields (Johnson and Keltner, 1983).

8.2.4.2 Useful Field of View (UFOV)

An observer's ability to detect, identify, and discriminate visual stimuli without eye movements is restricted to a relatively small region of the visual field known as the *useful field of view* (UFOV). The UFOV is reduced among older observers (Sekuler and Ball, 1986; Scialfa et al., 1987), due more to cognitive factors (e.g., attention) than basic sensory change. Ball and Owsley and their co-workers (Ball and Owsley, 1991; Ball et al., 1993; Sims et al., 2000) have shown significant reductions in UFOV to be predictive of involvement in "at-fault" automobile accidents. Owsley et al. (1998) found that older drivers with a UFOV impairment of 40% or greater were 2.2 times more likely to be involved in a car crash during the 3-year follow-up period of the study than those with little or no UFOV loss. Although not yet tested empirically, a reduction in UFOV may reduce older operators' ability to attend to and process road sign information.

8.2.5 Eye Movements and Visual Search

8.2.5.1 Aging Effects on Eye Movements

Aging appears to degrade the *smooth pursuit* eye movements that are used to follow targets in continuous motion relative to the observer (Zackon and Sharpe, 1987) as well as the ballistic *saccadic* eye movements employed to inspect different regions of the visual field. Older observers are usually slower to initiate saccadic eye movements (Munoz et al., 1998; Butler et al., 1999), have difficulty inhibiting reflexive saccades to irrelevant targets (Kramer et al., 2000), and have less accurate saccades (Butler et al., 1999). An age-related decrease in the range of ocular movement has also been reported (Clark and Isenberg, 2001). Such changes may erode older observers' ability to locate and process sign information, especially if it is in the peripheral field or in rapid motion.

8.2.5.2 Visual Search

Visual search tasks require observers to find a target stimulus within an array of distracter stimuli (e.g., looking for a specific sign against a background of many signs). Search time is extended by greater target/distracter similarity and by increased display size. Older adults, generally slower to locate targets, have particular difficulty when targets and distracters are similar, when many distracters are present (Scialfa and Joffe, 1997), or when the target shares one or more attributes with the distracters (Scialfa et al., 1998). Reductions in the latency and speed of saccadic eye movements appear to contribute significantly to older observers' slow search performance (Scialfa et al., 1994). These changes suggest that a sign located in a cluttered scene would be more difficult for older operators to locate; it is also possible that signs in such a location could contribute to older observers' increased failure to heed them (Huston and Janke, 1986; McGwin and Brown, 1999).

8.3 TRANSPORT SIGNS AND THE AGING OPERATOR

Because they are potential sources of critical information, traffic signs must be noticed, read, and understood. The associated time and cognitive effort for this can be limited, especially in conditions of high information load. For this reason, the appropriate response to a sign should be immediately clear, especially if the requisite action must be prompt (e.g., yield to approaching traffic). This issue is of increasing concern for older drivers as their response speed slows (Vercruyssen, 1997) and proneness to the effects of information overload and the distracting effects of irrelevant stimuli increase (Plude and Hoyer, 1986). Even signs developed in compliance with manuals on traffic control devices can be problematic. These include:

- Small print on word signs.
- Confusing word messages.
- Incomprehensible or illegible symbols.
- Low contrast between sign and background.
- Information overload (too much on one sign or a proliferation of signs in a limited area).
- Dirt on and/or deterioration of signs.
- Poorly placed or obscured signs.
- Out-of-date information (e.g., in work zones).
- Missing information or missing signs.

(See Chapter 3 and Chapter 4 for a systematic discussion of the criteria for determining sign effectiveness.)

8.3.1 Conspicuity

The degree to which a sign stands out sufficiently from its surroundings to capture attention (i.e., its *conspicuity*) is a function of several variables. These variables include sign size; angle of observation relative to the line of sight; color, brightness, and contrast relative to other objects in the surrounding area; and dynamic factors (change or movement), as well as such operator characteristics as alertness and search strategy (Mace, 1988).

Because old observers take longer to detect signs, especially in cluttered or visually complex environments, sign salience is more likely to be a challenge for them than for young observers. Dewar et al. (1994b) found that older drivers took about 34% longer than middle-aged drivers and about 70% longer than young drivers to locate a designated target symbol within an array of 18 different symbol signs. Ho et al. (2001) examined the effects of visual clutter, luminance, and driver age on the search conspicuity of traffic signs located within digitized traffic scenes. Compared to those of their young counterparts, the eye fixations of older drivers were slower and less accurate during the search process and older drivers took longer to determine if a target sign was present. Luminance was not a factor in sign detection.

The careful placement of signs, low visual clutter, and large sign size are especially likely to benefit older operators. When Dissanayake and Lu (2001) studied the effect of sign size on deceleration distance at STOP signs, they reported that older drivers began their deceleration at greater distances than did young and middle-aged drivers when large (48-in.) signs were tested.

8.3.2 Legibility

A sign must be legible from a distance that allows the operator sufficient time to respond appropriately. Sign legibility, usually measured in terms of either the minimum size (i.e., legibility size) or the maximum distance at which it can be identified (i.e., legibility distance), is a function of many factors. These include:

- Character size and separation.
- Brightness.
- Background luminance.
- Luminance and chromatic contrast between the legend and background.
- Visual complexity of symbols.
- Length and layout of word messages.
- Operator visual abilities.
- Familiarity with the sign and such environmental factors as weather (e.g., fog, snow, or rain) and lighting.

Licensing agencies have generally allowed persons with daylight acuities of 20/40 (6/12) or better to drive; in some jurisdictions the standard is even more lenient. The longstanding legibility guideline for word signs of 50 ft/in. (6 m/cm) of letter height, however, is equivalent to a 20/23 (6/6.5) acuity target. This can lead to a serious mismatch between allowable acuity and the legibility of the word signs to which drivers must respond. The recent adoption of a letter size guideline of 40 ft/in. (4.8 m/cm) in the U.S., equivalent to an acuity guideline of 20/29 (6/8.7), ameliorates this concern somewhat. For drivers with daytime acuity worse than 20/29, however, a lack of correspondence between letter size and acuity remains. This concern is amplified when acuity is further reduced by dusk or night lighting conditions, especially for older eyes.

Most prior research has shown sign legibility distances to be reduced among old observers. Sivak et al. (1981) examined the effects of age on the nighttime legibility of traffic signs composed of white letters on a dark background (green, red, blue, or black) or black letters on a light background (white, yellow, or orange). Legibility distances for drivers over 61 were only 65 to 77% those of young adult drivers with comparable high-luminance acuity. In a later field study, Sivak and Olson (1982) found that matching the young and old groups on low-luminance/high-contrast acuity eliminated the age difference in the nighttime legibility of signs.

Evans and Ginsburg (1985) compared young and old observers on the size of dynamically projected symbol signs that could be discriminated and found that legibility was about 30% greater for the young observers. Discrimination distance was related to contrast sensitivity at spatial frequencies of 1.5 and 12.0 c/deg but not to Snellen letter acuity. This result is consistent with Owsley and Sloane's finding (1987) that older persons need higher levels of contrast to detect and identify real-world targets, including signs.

Staplin et al. (1989) compared young/middle-aged (18 to 49 years) and old observers (65 to 80 years) on CS for a 20/80 Landolt C acuity target and letter size needed to read individual words and whole messages on novel four-word signs. Under nighttime conditions, the mean contrast levels needed by older drivers were about 2 to 2.5 times those of the young on the Landolt task; for the worst performers, the factor was as high as 20. Old observers required larger letters to read individual words as well as whole messages. Whereas variation of luminance had little effect on sign reading among the young, older observers benefited when luminance was increased. For both age groups, a white-on-green guide sign was the most difficult to read regardless of luminance, due perhaps to its low contrast. Contrast sensitivity was a relatively good predictor of word sign legibility, especially at lower luminance levels.

Allen et al. (1980) developed the following equation to estimate the effects of driver age on the recognition distance (in feet) of symbol signs:

$$\text{Recognition distance} = 262 - 2.23 \times \text{age} - 1.9 \times \text{errors}$$

Young adult drivers responded to signs at approximately (260 feet) 80 m and required approximately 2 sec of processing time; response distance deterioration with age was estimated to be 2.2 ft (.67 m) per year. In view of older drivers' relatively lower driving speeds, the authors concluded that "an additional sign recognition time penalty of 1.5 sec should be computed for drivers in their seventies."

In consideration of the limitations of older drivers, Mace (1988) refers to the concept of *minimum required visibility distance* (*MRVD*) — the distance from a sign at which drivers must detect and understand it, make a decision, and complete any needed vehicle maneuver. The MRVD is increased for older drivers for several reasons: slower detection time, increased time to understand unclear messages, diminished visual acuity and contrast sensitivity, and slower decision making. Older drivers also tend to have more difficulty when the roadway scene is visually complex, sign luminance is inadequate, or legend size is small.

8.3.2.1 Legibility: Symbol vs. Word Signs

Legibility distance is generally greater for symbol signs than their corresponding word version. Dewar and Ells (1974) found that young subjects driving toward signs could identify symbols at a mean of 262.4 m, a distance almost 60% greater than that for word signs. Jacobs et al. (1975) estimated the relative legibility distances of symbol and word signs for young observers under varied levels of optical blur. Without blur, the mean legibility distance of the symbol signs was almost twice that of word signs. When observer acuity was degraded by optical blur, the legibility advantage of symbol signs over words was even more pronounced.

Paniati (1989) found that the mean legibility distance of warning sign symbols was about 45% greater for young (20 to 45) than old (55 to 68) observers. The overall legibility of symbol signs was 2.8 times greater than the corresponding word versions, although the advantage of symbols varied widely for individual signs. The author noted that, for both age groups, legibility was best for "bold symbols of simple design." Kline et al. (1990) compared young, middle-aged, and elderly observers on the legibility distances of corresponding versions of four-word and four-symbol signs under simulated day and dusk lighting conditions. No age differences were seen on symbol comprehension or overall legibility. The mean legibility distance of the symbol signs, about double that of word signs for all three age groups, was greater in day than dusk conditions.

A systematic evaluation of the legibility of 85 symbol signs in the U.S. traffic control device manual (U.S. Department of Transportation, 1988) by Dewar et al. (1994b) found marked differences between signs and age groups. Mean overall legibility distance ranged from a high of 347.2 m (1146 ft) for the "Cross Road" symbol to a low of 44.4 m (147 ft) for "No Hitchhiking." Low legibility distance was associated with small details and/or inadequate separation between critical symbol features. Mean daytime legibility distances of middle-aged and old drivers were 88 and 80% those of young drivers, respectively. A significant legibility deficit was seen among old drivers on 19 of the 85 symbols tested.

Additional experiments evaluated the legibility distance of 18 additional "optimized" symbol signs in lighting that simulated daytime, nighttime, and a nighttime transient glare condition representative of oncoming headlights. Relative to daytime conditions, nighttime illumination degraded symbol legibility for all three age groups. However, the overall rank-order relationships between signs in day and night,

TABLE 8.1
Relative Legibility Distance for Symbol Signs for Different Drivers in Different Lighting Conditions

Age Group	Lighting Condition		
	Day	Night	Night/Glare
Young	1.00	0.70	0.69
Middle-aged	0.88	0.60	0.60
Old	0.80	0.64	0.34

Source: From Dewar, R.E. et al., 1994, *Transp. Res. Rec.*, 1456, 1–10.

as well as between day and night with glare for the same symbol signs, were near perfect (r = .97), indicating that effective legibility design transcends lighting condition. Relative to their nighttime legibility levels, young and middle-aged drivers were little affected by transient glare. Among old drivers, however, transient glare degraded nighttime symbol legibility. Symbol legibility for older drivers, about 80% that of young drivers in daylight, fell to about 50% with glare. The relative legibility distances for young, middle-aged, and old groups as a function of lighting condition are shown in Table 8.1.

Recent evidence suggests that older observers are better able than young ones to resolve optically degraded word signs. When Kline et al. (1999) reduced the acuity of all of their observers to 20/40 or 20/30, the old were better able than the young to identify the signs in day and dusk lighting conditions. A second experiment showed that older observers were also better able to identify optically degraded novel signs, indicating that their greater tolerance of blur is a generic ability rather than one based on greater familiarity with the blur profile of specific signs. A follow-up study (Bartel and Kline, 2002), however, showed that older observers were less able than young ones to identify faces, words, or scenes that were blurred intrinsically by low-pass digital image processing. This suggests that age-related changes in the optical media of the eye rather than "top-down" cognitive effects explain older observers' superior tolerance of optically blurred text.

8.3.3 Glance Legibility

Under certain driving conditions, drivers must obtain information from signs that are seen only briefly. The extent to which a sign meets this need is referred to as its *glance legibility*. Dewar et al. (1994b) compared the glance legibility of 18 briefly presented symbol signs for young, middle-aged, and old drivers. They found that the minimum duration needed to identify the critical details constituting each symbol rose with age: the mean threshold for old drivers was 2.2 times and 1.6 times that of the young and middle-aged drivers, respectively. The young/old group difference was significant for all 18 signs; the middle-aged/old difference was significant on only 3 signs.

8.3.4 Comprehension

Comprehension refers to the degree to which a driver can readily understand a sign's intended message. Kline et al. (1990) found considerable variability in the degree to which four symbol signs were understood by drivers of different ages. The symbol for Hill was understood by 85% of their participants; for Men Working by 75%, for Divided Highway by 60%, and for Road Narrows by only 52%. Although middle-aged drivers tended to understand the symbols (78%) better than young (62%) or elderly drivers (64%), the difference was not significant. A comparison of young, middle-aged, and old drivers by Dewar et al. (1994a), however, found an age-related deficit on the mean comprehension of 85 symbol signs in the U.S. uniform traffic control device manual. Older drivers' comprehension was worse than that of their younger counterparts on 33 (39%) of the symbols. Overall comprehension scores were below the 50% level for 13 of the 85 symbols.

Comprehension problems are not limited to symbol signs. When Hawkins et al. (1993) assessed comprehension of 15 word signs, they found eight (Speed Zone Ahead, High Occupancy Vehicle Restriction, Protected Left Turn on Green Arrow, Protected Left on Green, Lane Ends Merge Left, Grooved Pavement Ahead, Limited Sight Distance, and Ramp Metered When Flashing) that were understood by fewer than 67% of drivers.

8.3.5 Perception–Response Time

Perception–Response Time (PRT) refers to the time needed by drivers to perceive, analyze, and react to a traffic event. A sign's PRT is a function of "reading time" (the duration to fixate and read a sign and then return fixation to the road) and "decision time" (the interval needed to make a decision based on the sign information and to take the appropriate action). Reading time depends upon the length and complexity of the message; for a lengthy word sign, this may take several seconds and require two or more eye fixations.

PRT can be measured by displaying a sign's image and recording the time taken to identify its message or to indicate whether it matches a prior target message. Ells and Dewar (1979) had observers indicate verbally if a sign shown on a screen was the same as that presented vocally by the experimenter. They found that responses to symbol signs were about 17% faster than responses to words. Whitaker (1979) measured driver response times for directional messages conveyed by words, symbols (arrows), and symbol-plus-word combinations. Responses were faster to symbols than to words or to the symbol–word combinations.

Dewar et al. (1994b) measured the response times of young, middle-aged, and old drivers who determined if a symbol sign was the same as or different from the text name of the sign that had immediately preceded it on a computer screen. Response times (RTs) to identify the 18 symbols ranged from about 0.5 to 1.3 sec, depending on the sign and driver age. The mean RTs of middle-aged drivers (712 msec) were 29% longer and those of old drivers (930 msec) were 69% longer than those of young drivers (551 msec).

8.4 OPTIMIZING SIGN LEGIBILITY

8.4.1 Word Signs

Sign fonts vary somewhat in legibility. The Series E (Modified) font is used commonly for highway guide signs in the U.S. Clearview, a relatively new font designed to reduce the irradiation effects of retroreflective sign materials, has produced favorable results in testing. Garvey et al. (1997) compared the daytime legibility of Series D and Series E (Modified) alphabets with two versions of the Clearview alphabet: one normally sized and the other expanded so that the word's overall size was equal to that in the Series E (Modified) font among drivers 65 and older. Legibility distances were only marginally better for the Clearview font in daytime conditions, but at night the expanded Clearview font was better than the Series E (Modified) by 22% for type III sheeting and by 11% for type IX sheeting. Hawkins et al. (1999) found that Clearview was about 2 to 8% more legible than Series E (Modified) for overhead signs under daytime and nighttime viewing. For ground-mounted signs, however, Clearview was less legible in daytime and slightly more legible at night. In general, legibility improvement with Clearview was greatest for those with poor acuity — a benefit of particular relevance for aged drivers.

Carlson and Brinkmeyer (2002) compared the legibility distances of Clearview and Series E (Modified) fonts on overhead and shoulder-mounted freeway guide signs. Legibility was tested by having young (18 to 34 years), middle-aged (36 to 54 years), and older (55 years or older) drivers drive a vehicle toward signs under nighttime conditions. Overall, the Clearview font provided greater legibility distances than did the Series E (Modified) font. The average distance advantage of the former ranged from 3.1 to 9.4% for shoulder-mounted signs and from 3.2 to 8.6% for overhead signs. Although their legibility distances were shorter than those of the young and middle-aged, older drivers benefited the most from the Clearview font, with a 6.0% increase in legibility distance and thus reading time (e.g., 0.45 sec at a speed of 70 mph).

8.4.2 Symbol Signs

The legibility advantage of symbol signs over the corresponding word version gives the driver more time to determine and respond to their messages. Symbol signs can also be responded to more quickly under degraded viewing conditions (Ells and Dewar, 1979). These benefits appear attributable to the large (i.e., low spatial frequency) contours and contour separations of which symbols are usually composed. Conversely, as a symbol's details become finer (of higher spatial frequency), its legibility distance is reduced (Paniati, 1989; Kline et al. 1990). Kline and Fuchs (1993) determined the degree to which the legibility of the standard versions of symbol signs could be enhanced by optimizing the resistance of critical contours and contour separations to low-pass optical filtering. Four symbol signs were

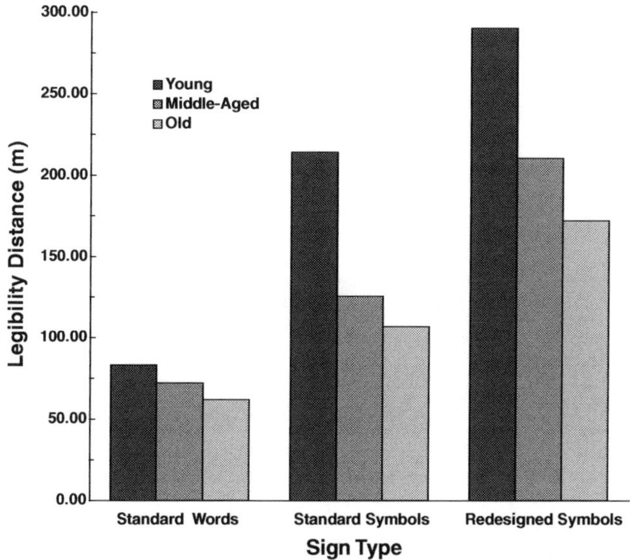

FIGURE 8.1 Legibility distance of standard text, standard symbol, and improved, redesigned symbol signs. (From Human Factors. Copyright 1993 by the Human Factors and Ergonomics Society. All rights reserved. Reprinted with permission: From the HFES.)

reiteratively viewed and redesigned through 5- to 7-D positive sphere ophthalmic lenses. The legibility distances of the redesigned symbols were then compared to those of the corresponding standard symbol and word versions among young, middle-aged, and elderly drivers.

Consistent with prior research, the mean legibility distance of the standard symbol signs was about double that of the word version for all three age groups. The legibility advantage of the redesigned symbols was even greater — about three times that of the standard word version and 50% greater than the standard symbol version (see Figure 8.1). The legibility gains among elderly drivers exceeded those of the young for three of the four redesigned symbols. For all three age groups, comprehension of the redesigned symbols was similar to that for the standard symbol version. Subsequent research (Long and Kearns, 1996) has shown that the legibility benefits of Kline and Fuchs' redesigned symbols are maintained under dynamic viewing conditions for all but the very fastest motion rates. These findings suggest that some of the legibility problems of older drivers could be alleviated by the use of well-designed symbol signs.

A computer-based, image-processing approach advanced by Kline et al. (Kline, 1991, 1994a, b; Schieber et al., 1994; Dewar et al., 1994a) has also shown considerable promise for optimizing the legibility of spatially complex irregular displays. Based on Fourier analytic techniques analogous to Kline and Fuchs' (1993) low-pass optical approach for enhancing legibility, a symbol is filtered through low-pass digital filters to identify and redesign any critical aspect of its message carried by high spatial frequency information. A recursive filtering/redesign process is then carried out until no further legibility gains can be realized. This technique has proven successful for enhancing legibility of a wide range of symbol signs in the U.S. sign system for young, middle-aged, and old drivers (Dewar et al., 1994b).

8.5 STRATEGIES FOR ENHANCING SIGNAGE FOR OLDER DRIVERS

Research on changes in vision and driving suggests a number of strategies for alleviating the problems that older operators experience in using sign information. Some of them are actions that can be taken by operators; others can be undertaken by traffic professionals, including automotive engineers, traffic engineers, and sign designers.

8.5.1 Strategies for Older Operators

No Human Factors solutions are yet available that can ameliorate fully the adverse effects of age-related sensory, cognitive, and psychomotor decline on the effectiveness of signs. However, a number of positive steps may be suggested for older operators and those who support them, including:

- Assure good general, sensory, and cognitive health through regular check-ups, diet, and exercise.
- Keep vision and hearing prescriptions up to date.
- Become informed about the effects of any medication on vision, cognitive function, motor skills, or driving.
- Clean eyeglasses, windshields, and headlights regularly.
- Select or add large side- and rear-view mirrors.
- Close the roadside eye briefly to avoid nighttime glare from oncoming headlights.
- Leave a bright flashlight in the vehicle for map and sign reading.
- Preview short nighttime trips during daylight hours.
- Preview road trips using maps.
- Travel with a navigator/companion.
- Minimize driving in high-traffic, nighttime, and poor-weather conditions.
- Consider purchasing a vehicle with a user-friendly, legible navigation system.
- Consider a driving course specialized for older operators.

8.5.2 Strategies for Traffic Professionals

In designing for older observers it is critical to recognize that variability on virtually all tasks increases markedly with age and that designs benefiting the least fit older person are likely to help drivers of all ages and skill levels. Successful system design for older drivers must be driven by a thorough understanding of the complex interactive sensory, cognitive, and psychomotor changes that affect driving. Specific suggestions include:

- Provide high-level, low-glare illumination on roadways, markers, and signs, especially at intersections.
- Employ bright, high-contrast materials for all traffic control devices, including lights, signs, and lane markers.
- Use large luminance contrast steps on signs, especially for those that require short-wavelength discriminations among blues and greens.
- Use large simple fonts and symbols with proven legibility.
- Enhance sign legibility by using optimized, well-comprehended symbols.
- Use optical or image processing techniques to optimize symbol and font legibility.
- Avoid visual clutter in the vicinity of signs.
- Tint only the top part of windshields and rear windows.
- Develop more powerful low-glare headlights and auto-dimming systems.
- Provide large side- and rear-view mirrors on vehicles.
- Develop simple, user-friendly navigation systems consistent with the needs of older operators.

8.6 CONCLUSIONS AND FUTURE RESEARCH

The problems of transport signs are frequently greater for older operators. Signs are more likely to be missed, not read in time, or misunderstood, and a sign's intended message may be confusing or misleading; symbols may be unclear; word messages can be ambiguous; and the print may be too small. In addition, signs are often placed in environments that make them difficult to detect. The increased likelihood of older operators' failing to heed signs may be due to a variety of factors, including failure to notice them, lack of understanding, inadequate legibility distance, or deliberate noncompliance. Research has yet to establish the full extent to which sign problems contribute to older drivers' elevated traffic accident rates. Some directions for potentially promising research are:

- Minimum sign legibility distances need to be established for different driving tasks as a function of driver age, health, visual capacity, skill, experience, and driving condition.
- Although there is an extensive and growing body of research on the effects of aging on automobile driving, little is known about age effects on other transport tasks (e.g., those associated with truck, bus, train, marine, and air transport).
- Age-appropriate perception–reaction time norms need to be developed for different transport tasks and incorporated into design guidelines for traffic control devices.
- Because most lab studies have investigated a limited number of variables, usually in the context of narrowly defined driving tasks, little is known about complex age-related interactive effects that might impair driving, such as those between multiple sensory deficits or health problems, or how such interactions might change in different driving conditions.

- Research in high-fidelity simulator systems offers considerable promise for understanding how to optimize sign conspicuity, legibility, and comprehension for realistic operating conditions.
- Many of the findings from lab and simulator studies need to be validated by field research.
- Relatively little research has been directed at the benefits and costs of older drivers' compensatory strategies for dealing with sign reading problems, their potential use by other drivers, and their possible utility for informing the design of more effective traffic control devices.

REFERENCES

Allen, R.W., Parseghain, Z., and Van Valkenburg, P.C.A. (1980). Simulator evaluation of age effects on symbol sign recognition. *Proc. Hum. Factors Soc. 24th Annu. Meet.*, Los Angeles (October).

Attebo, K., Mitchell, P., and Smith, W. (1996). Visual acuity and the causes of visual loss in Australia. *Ophthalmology*, 103(3), 357–364.

Ball, K. and Owsley, C. (1991). Identifying correlates of accident involvement for the older driver. *Hum. Factors*, 33, 583–595.

Ball, K., Owsley, C., Sloane, M.E., Roenker, D.L., and Bruni, J.R. (1993). Visual attention problems as a predictor of vehicle crashes in older drivers. *Invest. Ophthalmol. Visual Sci.*, 34(11), 3110–3123.

Bartel, P. and Kline, D.W. (2002). Aging effects on the identification of digitally blurred text, scenes and faces: evidence for optical compensation on everyday tasks in the senescent eye. *Ageing Int.*, 27, 56–72.

Brabyn, J., Schneck, M., Haegerstrom-Portnoy, G., and Lott, L. (2001). The Smith–Kettlewell Institute (SKI) longitudinal study of vision function and its impact among the elderly: an overview. *Optometry Vision Sci.*, 78(5), 264–269.

Bruckner, R. (1967). Longitudinal research on the eye (Basel Studies, 1955–1965), *Gerontol. Clin.*, 9, 87–95.

Burg, A. (1966). Visual acuity as measured by static and dynamic tests: a comparative evaluation. *J. Appl. Psychol.*, 50, 460–466.

Burns, N.R., Nettelbeck, T., White, M., and Willson, J. (1999). Effects of car window tinting on visual performance: a comparison of elderly and young drivers. *Ergonomics*, 42(3), 428–443.

Burns, P.C. (1999). Navigation and the mobility of older drivers. *J. Gerontol.: Psychol. Sci. Social Sci.*, 54(1), S49–S55.

Butler, K.M., Zacks, R.T., and Henderson, J.M. (1999). Suppression of reflexive saccades in younger and older adults: age comparisons on an antisaccade task. *Memory Cognit.*, 27(4), 584–591.

Carlson, P.J. and Brinkmeyer, G. (2002). Evaluation of Clearview on freeway guide signs with microscopic sheeting. *Transp. Res. Rec.*, 1801, 27–38.

Clark, R.A. and Isenberg, S.J. (2001). The range of ocular movements decreases with aging. *J. Aapos: Am. Assoc. Pediatric Ophthalmol. Strabismus*, 5(1), 26–30.

Collins, M.J., Brown, B., and Bowman, K.J. (1989). Peripheral visual acuity and age. *Ophthalmic Physiol. Opt.*, 9, 314–316.

Cooper, B.A., Ward, M., Gowland, C.A., and McIntosh, J.M. (1991). The use of the Lanthony New Color Test in determining the effects of aging on color vision. *J. Gerontol.: Psychol. Sci.*, 46, 320–324.

Crassini, B., Brown, B., and Bowman, K. (1988). Age related changes in contrast sensitivity in central and peripheral retina. *Perception*, 17, 315–332.

Desai, M., Pratt, L.A., Lentzner, H., and Robinson, K.N. (2001). Trends in vision and hearing among older Americans. *Aging Trends*, *2*. Hyattsville, MD: National Center for Health Statistics.

Dewar, R.E. and Ells, J.G. (1974). A comparison of three methods for evaluating traffic signs. *Transp. Res. Rec.*, 503, 38–47.

Dewar, R.E., Kline, D.W., and Swanson, H.A. (1994a). Age differences in the comprehension of traffic sign symbols. *Transp. Res. Rec.*, 1456, 1–10.

Dewar, R.E., Kline, D.W., Schieber, F., and Swanson, H.A. (1994b). Symbol signing design for older drivers. Final Report. Federal Highway Administration, Report #FHWA-RD-94-069, Washington, D.C.

Dissanayake, S. and Lu, J.J. (2001). Effect of larger stop signs on older drivers. Paper presented at the 80th Annu. Meet. Transp. Res. Bd. Meet., Washington, D.C.

Elliott, D.B. and Whitaker, D. (1991). Changes in macular function throughout adulthood. *Doc. Ophthalmol.*, 76, 251–259.

Elliott, D.B., Whitaker, D., and MacVeigh, D. (1990). Neural contribution to spatiotemporal contrast sensitivity decline in healthy ageing eyes. *Vision Res.*, 30, 541–547.

Elliott, D.B., Yang, K.C., and Whitaker, D. (1995). Visual acuity changes throughout adulthood in normal, healthy eyes: seeing beyond 6/6. *Optometry Vision Sci.*, 72(3), 186–191.

Ells, J.G. and Dewar, R.E. (1979). Rapid comprehension of verbal and symbol traffic sign messages. *Hum. Factors*, 21, 161–168.

Evans, D.W. and Ginsburg, A.P. (1985). Contrast sensitivity predicts age differences in highway-sign discriminability. *Hum. Factors*, 27, 637–642.

Garvey, P.M., Pietrucha, M.T., and Meeker, D. (1997). Effects of font and capitalization on legibility of guide signs. *Transp. Res. Rec.*, 1605, 73–79.

Gudmundsdottir, E., Jonasson, F., Jonsson, V., Stefansson E., Sasaki, H., and Sasaki, K. (2000). "With-the-rule" astigmatism is not the rule in the elderly. Reykjavik Eye Study: a population-based study of refraction and visual acuity in citizens of Reykjavik 50 years and older. *Acta Ophthalmol. Scand.*, 78(6), 642–646.

Haegerstrom-Portnoy, G., Schneck, M.E., and Brabyn, J.A. (1999). Seeing into old age: vision function beyond acuity. *Optometry Vision Sci.*, 76(3), 141–158.

Hawkins, H.G., Picha, D.L., Wooldridge, M.D., Greene, F.K., and Brinkermeyer, G. (1999). Performance comparison of three highway guide sign alphabets. *Transp. Res. Rec.*, 1692, 9–16.

Hawkins, H.G., Womack, K.N., and Mounce, J.M. (1993). Driver comprehension of regulatory signs, warning signs and pavement markings. *Transp. Res. Rec.*, 1403, 67–82.

Ho, G., Scialfa, C.T., Caird, J.K., and Graw, T. (2001). Visual search for traffic signs: the effects of clutter, luminance and aging. *Hum. Factors*, 43(2), 194–207.

Holland, C.A. and Rabbitt, P.M. (1992). People's awareness of their age-related sensory and cognitive deficits and the implications for road safety. *Appl. Cognit. Psychol.*, 6(3), 217–231.

Huston, R. and Janke, M. (1986). Senior driver facts (Tech. Report CAL-DMV-RSS-86-82). Sacramento, California: Department of Motor Vehicles.

Ivers, R.Q., Mitchell P., and Cumming, R.G. (2000). Visual function tests, eye disease and symptoms of visual disability: a population-based assessment. *Clin. Exp. Ophthalmol.*, 28(1), 41–47.

Jacobs, R.J., Johnston, A.W. and Cole, B.L. (1975). The visibility of alphabetic and symbol traffic signs. *Austr. Road Res.*, 5, 68–86.

Johnson, C.A. and Keltner, J.L. (1983). Incidence of visual field loss and its relationship to driving performance. *Arch. Ophthalmol.*, 101, 371–375.

Kalsi, M., Heron, G., and Charman, W.N. (2001). Changes in the static accommodation response with age. *Ophthalmic Physiol. Opt.*, 21, 77–84.

Klein, B.E., Klein, R., Lee, K.E., and Cruickshanks, K.J. (1999). Associations of performance-based and self-reported measures of visual function. The Beaver Dam Eye Study. *Ophthalmic Epidemiol.*, 6, 49–60.

Klein, R., Klein, B.E.K., Lee, K.E., Cruickshanks, K.J., and Chappell, R.J. (2001). Changes in visual acuity in a population over a 10-year period. *Ophthalmology*, 108(10), 1757–1766.

Kline, D. (1991). Visual aging and the visibility of highway signs. *Exp. Aging Res.*, 17(2), 80–81.

Kline, D. (1994a). Optimizing the visibility of displays for older operators. *Exp. Aging Res.*, 20, 11–23.

Kline, D.W. (1994b, January). A computer-based approach to testing and optimizing symbol signs for young, middle-aged and elderly drivers. Paper presented in Symp. Methods Measuring Operator Performance Behav. Annu. Meet. Transp. Res. Bd., Washington, D.C.

Kline, D.W., Buck, K., Sell, Y., Bolan, T., and Dewar, R.E. (1999). Older observers' tolerance of optical blur: age differences in the identification of defocused text signs. *Hum. Factors*, 41(3), 356–364.

Kline, D.W. and Fuchs, P. (1993). The visibility of symbolic highway signs can be increased among drivers of all ages. *Hum. Factors*, 35(1), 25–34.

Kline, D.W., Kline, T.J.B., Fozard, J.L., Kosnik, W., Schieber, F., and Sekuler, R. (1992). Vision, aging and driving: the problems of older drivers. *J. Gerontol.: Psychol. Sci.*, 47, 27–34.

Kline, D.W., Schieber, F., Abusamra, L.A., and Coyne, A. (1983). Age, the eye and the visual channels: contrast sensitivity and response speed. *J. Gerontol.*, 38, 211–216.

Kline, D.W. and Scialfa, C.T. (1997). Sensory and perceptual functioning: basic research and human factors implications. In A.D. Fisk and W.A. Rogers (Eds.), *Handbook of Human Factors and the Older Adult*, 327–328. San Diego: Academic Press.

Kline, T.J.B., Ghali, L.M., Kline, D.W., and Brown, S. (1990). Visibility distance of highway signs among young, middle-aged and elderly observers: icons are better than text. *Hum. Factors*, 32, 609–619.

Knoblauch, K., Saunders, F., Kusuda, M., Hynes, R., Podgor, M., Higgins, K.E., and de Monasterio, F.M. (1987). Age and illuminance effects in the Farnsworth–Munsell 100-hue test. *Appl. Opt.*, 26, 1441–1448.

Kosnik, W.D., Sekuler, R., and Kline, D.W. (1990). Self-reported problems of older drivers. *Hum. Factors*, 32, 597–608.

Kramer, A.F., Hahn, S., Irwin, D.E., and Theeuwes, J. (2000). Age differences in the control of looking behavior: do you know where your eyes have been? *Psychol. Sci.*, 11, 210–217.

Lee, K.E., Klein, B.E., and Klein R. (1999). Changes in refractive error over a 5-year interval in the Beaver Dam Eye Study. *Invest. Ophthalmol. Visual Sci.*, 40(8), 1645–1649.

Long, G.M. and Crambert, R.F. (1990). The nature and basis of age-related changes in dynamic visual acuity. *Psychol. Aging*, 5, 138–143.

Long, G.M. and Kearns, D.F. (1996). Visibility of text and icon highway signs under dynamic viewing conditions. *Hum. Factors*, 38(4), 690–701.

Lyman, J.M., McGwin, G., Jr., and Sims, R.V. (2001). Factors related to driving difficulty and habits in older drivers. *Accident Anal. Prev.*, 33(3), 413–421.

Mace, D.J. (1988). Sign legibility and conspicuity. In *Transportation in an Aging Society: Improving Safety and Mobility for Older Persons, Vol. 2*, Washington D.C.: Transportation Research Board, 270–293.

Marottoli, R.A., Richardson, E.D., Stowe, M.H., Miller E.G., Brass L.M., Cooney, L.M., Jr., and Tinetti, M.E. (1998). Development of a test battery to identify older drivers at risk for self-reported adverse driving events. *J. Am. Geriatr. Soc.*, 46(5), 562–568.

McGwin, G., Jr. and Brown, D.B. (1999). Characteristics of traffic crashes among young, middle-aged and older drivers. *Accident Anal. Prev.*, 31(3), 181–198.

McKnight, A.J. and McKnight, A.S. (1999). Multivariate analysis of age-related driver ability and performance deficits. *Accident Anal. Prev.*, 31(5), 445–454.

Munoz, D.P., Broughton, J.R., Goldring, J.E., and Armstrong, I.T. (1998). Age-related performance of human subjects on saccadic eye movement tasks. *Exp. Brain Res.*, 121(4), 391–400.

Olson, P.L. and Sivak, M. (1984). Glare from automobile rear-vision mirrors. *Hum. Factors*, 26, 269–282.

Owsley C., McGwin, G., Jr., and Ball, K. (1998). Vision impairment, eye disease and injurious motor vehicle crashes in the elderly. *Ophthalmic Epidemiol.*, 5(2), 101–113.

Owsley, C. and Sloane, M.E. (1987). Contrast sensitivity, acuity and the perception of "real-world" targets. *Br. J. Ophthalmol.*, 71, 791–796.

Owsley, C., Sekuler, R., and Siemsen, D. (1983). Contrast sensitivity throughout adulthood. *Vision Res.*, 23, 689–699.

Paniati, J.F. (1989). Redesign and evaluation of selected work zone symbols. *Transp. Res. Rec.*, 1213, 47–55.

Plude, D.J. and Hoyer, W.J. (1986). Age and the selectivity of visual information processing. *Psychol. Aging*, 1, 1–16.

Reading, V.M. (1972). Visual resolution as measured by dynamic and static tests. *Pflugers Arch.*, 333, 17–26.

Ryan, G.A., Legge, M., and Rosman, D. (1998). Age-related changes in drivers' crash risk and crash type. *Accident Anal. Prev.*, 30(3), 379–387.

Schieber, F., Kline, D.W., and Dewar, R.E. (1994). Optimizing symbol highway signs for older drivers. *Proc. 12th Triennial Conf. Int. Ergonomics Assoc.*, 6, 199–201.

Schieber, F., Kline, D.W., Kline, T.J.B., and Fozard, J.L. (1992). The relationship between contrast sensitivity and the visual problems of older drivers. SAE Technical Paper Series, No. 920613, 1–7.

Scialfa, C.T., Adams, E.M., and Giovanetto, M. (1991). Reliability of the Vistech Contrast Test System in a life-span adult sample. *Optometry Vision Sci.*, 66, 270–274.

Scialfa, C.T., Esau, S.P., and Joffe, K.M. (1998). Age, target-distracter similarity and visual search. *Exp. Aging Res.*, 24, 337–358.

Scialfa, C.T., Garvey, P.M., Tyrrell, R.A., and Leibowitz, H.W. (1992). Age differences in dynamic contrast thresholds. *J. Gerontol.: Psychol. Sci.*, 47, 172–175.

Scialfa, C.T. and Joffe, K.M. (1997). Age differences in feature and conjunction search: implications for theories of visual search and generalized slowing. *Aging, Neuropsychol. Cognit.*, 4, 227–246.

Scialfa, C.T., Kline, D.W., and Lyman, B.J. (1987). Age differences in target identification as a function of retinal location and noise level: an examination of the useful field of view. *Psychol. Aging*, 2, 14–19.

Scialfa, C.T., Thomas, D.M., and Joffe, K.M. (1994). Age differences in the useful field of view: an eye movement analysis. *Optometry Vision Sci.*, 71, 736–742.

Sekuler, R. and Ball, K. (1986). Visual localization: age and practice. *J. Opt. Soc. Am. A*, 3, 864–867.

Shinar, D. and Schieber, F. (1991). Visual requirements for safety and mobility of older drivers. *Hum. Factors*, 33(5), 507–519.

Sims, R.V., McGwin, G., Jr., Allman, R.M., Ball, K., and Owsley, C. (2000). Exploratory study of incident vehicle crashes among older drivers. *J. Gerontol.: Med. Sci.*, *55A*, M22–M27.

Sivak, M. and Olson, L. (1982). Nighttime legibility of traffic signs: conditions eliminating the effects of driver age and disability glare. *Accident Anal. Prev.*, 14(2), 87–93.

Sivak, M., Olson, P., and Pastalan, L.A. (1981). Effects of driver age on nighttime legibility of traffic signs. *Hum. Factors*, 23, 59–64.

Stamatiadis, N. and Deacon, J.A. (1995). Trends in highway safety: effects of an aging population on accident propensity. *Accident Anal. Prev.*, 27(4), 443–459.

Staplin, L., Lococo, K., Sim, J., and Drapcho, M. (1989). Age differences in a visual information processing capability underlying traffic control device usage. *Transp. Res. Rec.*, 1244, 63–72.

Sturr, J.F., Kline, G.E., and Taub, H.A. (1990). Performance of young and older drivers on a static acuity test under photopic and mesopic luminance conditions. *Hum. Factors*, 32, 1–8.

U.S. Department of Transportation, Federal Highway Administration (1988). *Manual on Uniform Traffic Control Devices for Streets and Highways*. Washington, D.C.: Author.

Vercruyssen, M. (1997). Movement control and speed of behavior. In A.D. Fisk and W.A. Rogers (Eds.), *Handbook of Human Factors and the Older Adult*, 55–86. San Diego: Academic Press.

Whitaker, L. (1979). The effects of words vs. symbols on reaction times to left–right traffic signs. *Ergonomics*, 22, 765.

Yee, D. (1985). A survey of the traffic safety needs and problems of drivers age 55 and over. In J.L. Malfetti (Ed.), *Drivers 55±: Needs and Problems of Older Drivers: Survey Results and Recommendations*, 96–128. Falls Church, VA: AAA Foundation for Traffic Safety.

Zackon, D.H. and Sharpe, J.A. (1987). Smooth pursuit in senescence: effects of target acceleration and velocity. *Acta Otolaryngol.*, 104, 290–297.

9 Motivational Aspects of Traffic Signs

Ray Fuller

CONTENTS

9.1 Learning and Motivation 136
9.1.1 Traffic Signs as Discriminative Stimuli 138
9.1.2 Sign Effectiveness 141
9.1.3 Why Signs Should Represent Contingencies Reliably 141
9.1.4 Using Signs to Reduce Uncertainty 146
9.2 Detecting Traffic Signs 148
9.3 Conclusions and Practical Implications 151
References 151

The road and traffic environment is a regulated system of which traffic signs are an essential and pervasive part. In an urban environment, Summala and Naatanen (1974) observed approximately 14 traffic signs per kilometer (i.e., on average 1.4 signs every 100 m), so the term *pervasive* is hardly an overstatement. As described elsewhere in this book, signs may be classed as *mandatory* (directing what must or must not be done), *warning* (advising of a potential hazard ahead), and *directional* (helping find and follow the desired road). To do their job, traffic signs must give their messages clearly and allow enough time for the driver to see them, understand them, and act appropriately (U.K. Driving Standards Agency, 1992).

When used successfully, traffic signs should enable a more effective, efficient, and safe use of the roadway. Thus, one might suppose that drivers would be keen to detect all signs. That this is demonstrably not the case is evidenced by research that stopped cars and questioned drivers about the signs they had just passed (e.g., Johansson and Rumar, 1966; Johansson and Backlund, 1970). Typically, only 47% of drivers could recall the signs. Further studies, which tested recall of signs "just passed," found proportions correct varying from as low as less than 10% (Milosevic and Gajic, 1986; Shinar and Drory, 1983) to 78% (see, for example, the summary of evidence by Lajunen et al., 1996). One might also suppose that drivers would be highly motivated to comply with the road signs they were able to detect. However, the universal observation that even when they see a sign, drivers may not respond to it (for example, the speed limit sign; Hagemeister and Westhoff, 1997) violates

0-415-31086-5/04/$0.00+$1.50

this supposition. This chapter, therefore, is concerned with the vital issue of motivation with respect to traffic signs.

9.1 LEARNING AND MOTIVATION

To understand why signs might fail in their intended functions, it is necessary to explore how signs operate from a psychological perspective. To do this, some basic principles concerning human learning, motivation, and attention are necessary. Fortunately, one conceptual model in psychology provides an accessible and powerful account of key aspects of all three, namely, behavior theory and its associated methodology of behavior analysis. The concepts of behavior theory have a long history, originating in the work of Watson, Thorndike, and Skinner in the first half of the 20th century. These pioneers were able to describe in a systematic way principles of learning and motivation that had driven human behavior throughout human evolution; these principles also drive elements of the behavior of many other animal species today that show behavioral plasticity, i.e., the ability to modify their behavior as a result of experience.

At the heart of behavior theory is the notion that behavior followed by a rewarding consequence becomes strengthened in the individual's repertoire of behaviors — a process known as reinforcement. The corollary to this is that behavior followed by an aversive consequence is weakened in the repertoire — a process known as punishment. Thus, things that lead to a pleasant consequence are more likely to be done in the future; things that have an unpleasant consequence are likely to be avoided in the future. One warms the hands in front of a fire on a cold night but does not put them into the flames.

Thus far, two terms are in this simple learning model, with a contingent relationship between them: behavior and its consequences. To complete it, one further term needs to be added — an element that represents conditions under which particular behavior–consequence contingencies work. After all, the rewarding consequence of approaching the fire to warm one's hands only happens if the fire is giving out heat. If it is a barely smoldering lump of coal or wood, one is hardly likely to approach the fire for warmth.

This antecedent condition, which sets the scene for the act of warming the hands to be rewarding (i.e., a fire with flames or a warm glow), is known as a discriminative stimulus. In effect, it tells an individual that placing his hands nearer the fire will now lead to a desired consequence. Through the same kind of process, very close proximity to the flames or a brightly glowing ember acts as a further discriminative stimulus that says in effect, "Come too close to me and you will be punished!" Thus, although someone puts his hands nearer the fire to warm them, he stops short of actually putting them into the fire.

Figure 9.1 is a representation of this three-term contingency model. Conveniently, the terms antecedent, behavior, and consequence yield the abbreviation ABC, providing a basic model to explain how one learns to emit particular behaviors under particular circumstances: the ABC of behavior. Table 9.1 gives further examples of antecedent events (discriminative stimuli), behaviors, and consequences.

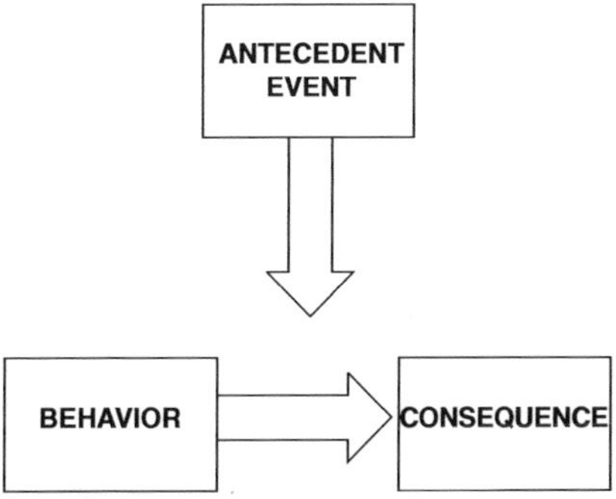

FIGURE 9.1 The three-term contingency model of behavior theory.

TABLE 9.1
Examples of Antecedent Events (Discriminative Stimuli), Behaviors, and Consequences

Discriminative Stimulus	Behavior	Consequence
Glowing fire	Approach	Feel warmer
Phone rings	Pick up phone	Conversation
Headache	Take painkiller	Pain suppressed
Room in darkness	Switch on light	See what one is doing
Feeling too hot	Take off jacket	Feel more comfortable

Note that the first time a person approaches a fire (or does anything, for that matter), he does not need to know consequences of the behavior. Of course, in the example given, he will very quickly feel the heat and, if he goes too far, the pain of burning. Such learning will require few training experiences. However, once he has learned a contingent relationship, such as that exemplified earlier, he can represent it internally in his head so that the next time he wants to warm his hands he can approach the glowing fire *in order to obtain* the warming benefit.

Because an individual can mentally represent consequences and because experience teaches him to expect those consequences, behavior is referred to as goal directed and intentional. The consequences of behavior provide the motivation for it. People are motivated to approach the glowing fire because they anticipate the consequence of feeling warmer. In the same way, they are motivated to go to work because the consequence is the reward of payment. (Work may, of course, also have other rewards such as the intrinsic rewards of job satisfaction, social interaction, and power over others). With this simple three-term contingency model in mind, it is now possible to explore how driver behavior might be controlled on the roadway, how traffic signs work, and why signs might fail to work.

Imagine a long, fast, straight run approaching a sharp left-hand bend as in Figure 9.2. The stimulus of the disappearing road acts as a powerful discriminative stimulus that signals the contingency "high speed may be followed by loss of control of the

FIGURE 9.2 The natural environment provides a discriminative stimulus for behavioral contingencies.

vehicle, running off the roadway and/or collision with an as yet unseen stationary obstruction" — or, in other words, the behavior of high speed has a high probability of being punished. The corollary to this is that the disappearing road as discriminative stimulus signals the contingency (among others; see later in this chapter) that "slowing down will be followed by safe negotiation of this segment of roadway." In other words, the behavior of reducing speed will be rewarded. In the absence of traffic signs or road markings, people must rely on the available discriminative stimuli in the natural environment to select behaviors that will enable them to achieve their goals. In driving, these goals are presumably mobility with safety.

9.1.1 Traffic Signs as Discriminative Stimuli

Where does the traffic sign fit into this analysis? First, consider the warning sign, which provides an excellent example of a discriminative stimulus. It indicates (usually in a rather generalized and stylized way) the nature of the real-world, natural-environment discriminative stimulus ahead. In this way it acts as a surrogate for that natural environment and, like the natural environment, tells an individual the relationship between a particular behavior and its consequences. Consider, for example, the warning sign for a bend in the road ahead (see Figure 9.3).

This not only tells the kind of natural discriminative stimulus to be experienced ahead but, in effect, also tells that certain behaviors in the next segment of roadway will be punished, whereas others will avoid that punishment. It says that high speed may lead to loss of control of the vehicle and/or possible collision with a (at present) hidden obstruction. Slowing down will avoid these punishing consequences and be

FIGURE 9.3 Example of a sign warning of a bend ahead.

rewarded with a safe (and more comfortable) outcome. More than this, however, the warning sign also says that a low level of vigilance (active attention and search for possible threats) may be punished, whereas a high level is likely to be rewarded by enabling detection of an impending hazard. The advance nature of the warning sign not only suggests contingencies that might be expected but also prepares one to be ready for those contingencies, thereby enabling a quicker response if the need arises.

Warning signs that suggest a lower speed–safer consequence contingency may trigger only small changes in speed in the region immediately past the sign (see, for example, Summala and Hietamaki, 1984), but vigilance and readiness to respond to the real-world discriminative stimulus for danger will be enhanced. Finally, by preceding the actual hazard signified, the sign provides more time to make appropriate adjustments — for example, to your speed (Crundall and Underwood, 2002). Thus, a warning sign:

- Acts as a discriminative stimulus for behavior–consequence contingencies.
- May warn of a hazard that is not obvious from the naturally occurring discriminative stimuli.
- Enables a state of readiness to respond.
- Provides more time to adjust to the hazard ahead.

This same analysis applies equally well to mandatory and directional signs. From the first category, consider a stop sign located at a junction with a road of higher priority (see Figure 9.4). This discriminative stimulus says that proceeding into the junction without stopping will (possibly) be punished with collision with a vehicle on the higher-priority road. Because the sign is mandatory, it also tells an individual that violation of the requirement to stop may be further punished by a law enforcement agency's stopping him in his journey, "ticking him off," fining him, or taking further proceedings against him (in order of severity of consequence). Directional signs, on the other hand, provide a discriminative stimulus for route choice (see Figure 9.5). If one wishes to go to Newcastle, carrying straight on or taking the second turn left will not be rewarded by reaching the destination (it will also incur the penalty of a waste of time and fuel); only turning left immediately will be rewarded. At the same time, the sign says to look out for the junction needed and to be ready to take action.

FIGURE 9.4 A mandatory stop sign.

FIGURE 9.5 A directional sign.

9.1.2 Sign Effectiveness

For a traffic sign to be effective as a surrogate discriminative stimulus, its relationship with the natural environment must be learned. In other words, it is necessary to learn what each different sign represents; assessing that learning is a traditional feature of driver license testing in many jurisdictions. Not surprisingly, as experience of signs and the conditions they represent develops, an individual becomes more accurate at recognizing the signs. As is mentioned elsewhere in this book, Al-Madani and Al-Janahi (2002), using a multiple-choice sign recognition test, found that male drivers with more than 10 years' driving experience were significantly better at sign recognition than less experienced male drivers. With experience one also becomes quicker at recognizing the conditions represented by a sign after seeing it. In a laboratory study, Crundall and Underwood (2002) demonstrated that among inexperienced drivers, a difference of only 3 years' driving experience produced an average reduction of 179 ms in the time needed to recognize the conditions represented by three different warning signs.

Before recognition can take place, however, traffic signs need to be detected by the approaching driver. They also need to signal in a reliable way the contingency between particular behaviors and their consequences. These two issues will be covered in reverse order. It will be shown first that unreliable contingencies between the behavior advised by a sign and its consequences can seriously undermine sign effectiveness. Later it will be shown that design features of signs, their locations, and the amount of information given can come up against limitations of human information processing.

9.1.3 Why Signs Should Represent Contingencies Reliably

As discussed earlier, a discriminative stimulus works because it signals the relationship between a particular behavior and its consequences. However, if that relationship is variable — sometimes it happens and sometimes it does not — then it is difficult for the discriminative stimulus to become established in the first place. One of the warning signs that attracted attention from this perspective many years ago was that for "Road Work Ahead." A modern example is shown in Figure 9.6.

Gardner and Rockwell (1983) found that drivers passing such signs typically failed to change their behavior and only prepared to slow their speed if they could actually see an obstruction in the roadway ahead. Interviews with drivers revealed that they had often passed warning signs left at the roadside long after the roadwork had been completed and the contractors departed. Clearly, such signs were no longer signaling the contingency between approach speed and a possible collision with, for example, slow-moving construction equipment; this learning had generalized to other instances of the warning sign.

Another example is the "Children Crossing" sign, most typically located at the approach to a school (see Figure 9.7). Research in the U.K. by Howarth (1988) revealed that drivers passing this type of sign did not reduce speed; in fact, they maintained speed even when children were observable on the footpath. A fundamental problem with this sign is that the contingency it signals (possible collision

FIGURE 9.6 A warning sign for roadwork ahead.

with a child dashing across the road or a group of children being shepherded across by a crossing guard, for example) really applies only when the children are in transit to and from school. Thus, for most of each day and, indeed, for the periods each year when schools are not in session, the sign is irrelevant. Is it any surprise, then, that the contingency the sign is meant to represent fails to control driver behavior? The obvious way of overcoming this problem is to make the relationship between the sign and the contingency it is intended to represent more reliable. This has been done in several variations of the sign, one of which is presented in Figure 9.8. With this sign design, alternately flashing amber lights are switched on only for those periods in which the child is in transition to and from school.

The sign that perhaps most frequently fails to achieve its function is the mandatory sign for a speed limit. As with other mandatory signs, it acts as a discriminative stimulus for two kinds of punishing event: comply or risk punishment by losing control of the vehicle or a collision (or both) and comply or be liable to penalty under the law. Warning signs typically tell about something that might be discovered on the roadway, anyway; mandatory signs say something that might not otherwise be known. Thus, perhaps not surprisingly, drivers adjust speed more to a mandatory speed sign than to a warning sign appearing at different times in the same location (Summala and Hietamaki, 1984); they are also more likely to be able to recall a speed limit sign they have just passed (Johansson and Rumar, 1966).

However, consider what happens in practice when a driver violates a mandatory speed restriction. Imagine a driver entering a road with a signed speed limit of 110 km/h. What happens if the driver exceeds the limit by 10 km/h? Most times the driver simply gets to his destination in a little less time. The driver does not lose

FIGURE 9.7 A sign warning of children crossing.

FIGURE 9.8 Sign for children crossing that signals a reliable contingency.

FIGURE 9.9 Sign for the contingency of excessive speed and police-enforced penalty.

control of the vehicle or get involved in a collision and no penalty is imposed by the police. This is because the design speed of the road is possibly at least 20 km/h higher than the posted limit so can be safely traveled (o.t.r.e.) at the higher speed and because the agents of the law cannot be everywhere at once.

Reviewing relevant evidence, Wilde (1994, p. 187) reports that in one study in Montreal the ratio of incidents of speeding 10 km/h over the posted limit relative to the number of speeding charges was in the order of 7000:1. Thus, the driver who wishes to shorten the duration of a journey learns that, by and large, the speed limit can be exceeded with impunity. To the driver motivated to violate it, the speed limit sign acts less to signal the contingency between higher speed and danger and more as a signal to increase vigilance for a possible speed trap. Consequently, the presence of a police car strengthens the response of drivers to a speed restriction sign (Lajunen et al., 1996), and some jurisdictions have conceded this by introducing a sign that actually specifies the contingency of police-enforced penalty for exceeding the limit (see Figure 9.9).

Holland and Conner (1996) have demonstrated that this type of penalty warning sign can increase compliance by about 36% and that the addition of police presence further increases compliance by about 45% of preintervention levels. In dealing with the speed violation problem, technology has come to the aid of the enforcement agencies in the form of fixed and mobile speed cameras, on-board tachographs, "black" boxes, and individual vehicle speed capture using satellite monitoring. Thus, it is now possible for the enforced penalty contingency to be a reliable outcome for noncompliant behavior. Extensive application of this technology could have a remarkable effect in the future on the discriminative stimulus value of the speed restriction sign.

In urban and residential areas, however, where speed restrictions permit only much lower speeds (typically 40 or 50 km/h), the difference between design and

FIGURE 9.10 A consequence of exceeding the posted speed limit.

posted speed limits may be so much less that it is virtually nonexistent. Thus, exceeding the limit may indeed lead to loss of control of the vehicle (see Figure 9.10) or collision with another road user.

Nevertheless, at times it is possible to exceed the limit, undermining the discriminative stimulus power of the speed restriction sign. From the perspective of behavior analysis, what this is saying is that once the driver is past the speed restriction sign, noncompliant driver behavior will be largely controlled by the available discriminative stimuli in the environment. If those stimuli suggest that a higher speed will not be punished, the driver may well then exceed the speed limit. As has been mentioned earlier in this book, engineers have employed strategies to deal with this problem such as repeating the speed limit sign at intervals or employing a "reminder" sign, such as the "rappel" sign often used in France. Milosevic and Gajic (1986) have shown that simply repeating a sign increases the likelihood that it will be recalled by the driver when stopped a little further along the road.

In other instances, engineers have countered speed violations by introducing design changes collectively known as traffic calming measures. Examples include lane narrowing, chicanes, throats, and humps (see Figure 9.11). These work by punishing attempts to travel through them at an excessive speed; in other words, the link between the traffic sign and the violation–consequence contingency is considerably strengthened. The corresponding physical changes in the appearance of the roadway also, of course, provide different discriminative stimuli from those that prevailed before the road modification.

In the context of mandatory signs, the discriminative value of the sign to the road user in enabling avoidance of enforced penalties is clearly of value. The road and traffic system requires compliance for safety and optimized mobility, and it is preferable to induce this with appropriate discriminative stimuli than to have people learn to comply through direct experience of punishment. Powerful discriminative stimuli (such as the presence of a marked police vehicle) can even be simulated by two-dimensional fabric mockups stretched over a frame (see Figure 9.12).

FIGURE 9.11 Traffic-calming measures.

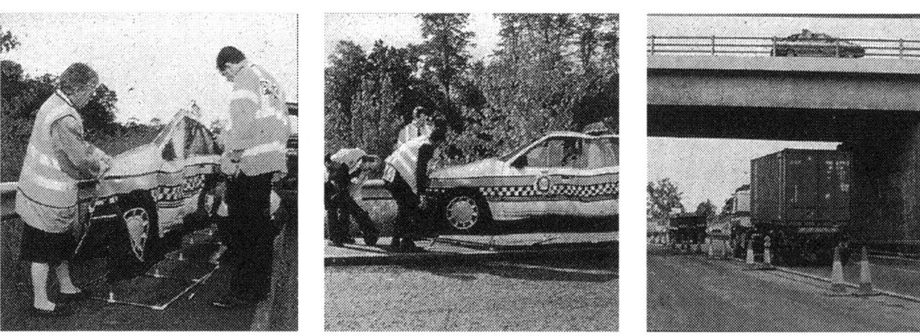

FIGURE 9.12 A "virtual" discriminative stimulus for the noncompliance–punishment contingency. (Courtesy of the Transport Research Laboratory, U.K.)

9.1.4 Using Signs to Reduce Uncertainty

Traveling over unfamiliar roads provides the driver with a challenge because he does not know the precise contingencies operating in each road segment. For example, on approach to an unfamiliar bend, what will be a safe maximum speed? Under such circumstances and with no other information, the driver must fall back on the available discriminative stimuli. How far can one see around the bend? Are there other clues to its radius (such as speed of vehicles coming out of the bend and approaching the driver)? What are the road surface conditions? Given this discriminative stimulus processing, the driver then must select a speed that past experience has taught will enable proceeding through the bend without loss of control. In effect, the discriminative stimuli select the equivalent of a conditional rule, which might be formalized as: "for a bend of this apparent radius and road surface, a speed of x km/h will deliver a safe outcome." In other words, learned contingencies between different bends and different speeds are generalized to the new situation.

This kind of conditional rule need not be articulated verbally or even represented in working memory; in the main, the driver must adjust speed so that the pattern of visual feedback, of *g* forces, and of steering wheel resistance matches some internally represented model of what an acceptable pattern feels like. Of course, the discriminative stimulus information may be inadequate to enable selection of an appropriate speed, as might the driver's previous learning experience. If the error means a slower than necessary speed is selected, then all that suffers, very slightly, is mobility. However, if the speed is too high, the consequence may be loss of control, unless conditions enable recovery from the error.

Traffic signs that specify what the driver must do (to avoid a punishing consequence) may be implemented to deal with this kind of "unfamiliarity" problem. A good example is the use of advisory or mandatory speed signs on bends and curves. Another example is to provide written instructions on the roadway, such as SLOW (see Figure 9.13).

Telling the driver what to do avoids the driver's having to discover the contingencies operating through guesswork (i.e., through trial and error) or through reliance on appropriate generalization from previous experience. Thus, the incidence of dangerously close headways can be decreased through use of a monitoring system measuring a driver's time headway and illuminating a warning sign if the driver is following too closely (Helliar–Symons, 1983). Marks on the roadway indicating safe following distances may also be used on high-speed roads such as highways where stream speed is reasonably predictable.

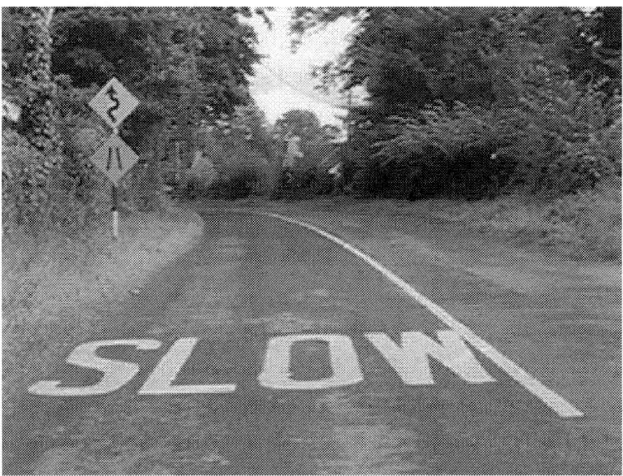

FIGURE 9.13 Direct instructions to the driver help avoid reliance on available discriminative stimuli and generalization of learned contingencies.

9.2 DETECTING TRAFFIC SIGNS

When no traffic signs are present, the stimulus control of behavior on the roadway is provided by the available discriminative stimuli in the physical environment. The traffic sign provides an *additional* discriminative stimulus, which takes some of the uncertainty out of the behavior–consequence contingency that might otherwise prevail. However, in many instances the additional information provided by the sign will essentially be redundant (see Macdonald and Hoffmann, 1991, for a useful classification).

Thus, it is not surprising that, in many studies in which drivers are stopped after passing a sign and asked to recall it (e.g., Milosevic and Gajic, 1986; Shinar and Drory, 1983; Lajunen et al., 1996), as few as 10% of drivers are able to do so. Similarly, in a procedure in which drivers had to report immediately whatever they were aware of that was relevant to the driving task 3 to 4 sec before their vision was suddenly occluded (they were in a dual-control vehicle), Macdonald and Hoffmann (1991) found that drivers were unaware of the information presented by most traffic signs. Furthermore, because the additional information provided by a sign may be redundant, it should also not be surprising that results of studies suggest that, even without conscious recall of signs, driver behavior is nevertheless modified (Summala and Hietamaki, 1984).

Directional signs, however, are a powerful example of how signs can remove uncertainty about behavior–response consequences; without them a driver might need to go a long way down an uncertain route before he is rewarded by approaching the desired destination or punished by discovering he has taken the wrong road. The same point applies as much to mandatory as to warning signs. They help to ensure that the appropriate behavioral choice is made in negotiating the next section of roadway.

Of course, for the traffic sign to achieve this effect, it must be detected, read, and understood; also, apart from all of these processes, enough time must be left to make whatever behavioral adjustment is required (such as stopping at a junction). Because speed of approach to a sign determines the time window of opportunity for all of these things to happen, approach speed becomes an important determinant of sign size characteristics and its positioning relative to the hazard, route choice, or mandatory requirement signified. It is perhaps not surprising then that one of the earliest strands of systematic human factors research on traffic signs was to determine legibility distances for standardized highway signs (Forbes, 1972).

In 1939, Forbes and Holmes published legibility distance data for black-on-white letters indicating destinations in daylight. Their results revealed a linear relationship between visibility distance and letter height of about 600:1 (50 ft:1 in.). Narrower letters gave a relationship of about 400:1. These results represented 80-percentile values for participants with normal 20/20 vision. When floodlit at night, legibility distance for these signs was reduced by between 10 and 20%. Subsequent research over the years by Forbes and many others has determined values for destination signs with different letter styles and cases, heights, widths, height-to-width ratios, spacing, brightness and color contrast within the sign, illumination, and reflectivity. Forbes (1972) reported that brightness contrast between the sign and its background contributes to a sign's ability to capture attention (replicated more recently by Schieber and Goodspeed, 1997, and Ho et al., 2001) and that background contrast and brightness contrast *within* the sign contribute independently to relative sign visibility (which signs are seen first).

Attention, however, is not a passive process. The disposition to attend to an external stimulus is not simply a passive state, waiting to be switched on by the most alluring stimulus, although, as will be shown in a moment, stimuli can differ quite substantially in their ability to capture attention. Attention is selective and may be conceptualized as a process that determines which incoming sensory stimuli get passed along to higher levels of perceptual processing. This is by no means a simple process. Stuss et al. (1995) identify seven aspects of attention: sustaining, concentrating, suppressing, switching, sharing, setting, and concentration. Groeger (2000, p. 56) has nicely indicated how all of these can be concurrently deployed during driving.

Furthermore, factors such as motivation, needs, and expectations may predispose a driver to attend to one stimulus rather than another. In contrast to relatively poor performance when drivers are not expecting to need to recall signs just passed, when asked to look out for signs on a journey, they are able to spot almost all of them (see, for example, Summala and Naatanen, 1974). In the same way, when there is a need to detect particular signs, such as in route finding and approaching intersections on a highway, destination and route signs may have a high priority for attention. Macdonald and Hoffmann (1991) found that the most important factor affecting level of reported sign information in their study was what they termed the sign's "action potential" — the rated probability of the driver's needing to make an overt response related to the sign's information. The strength of this relationship implies that most signs must be initially processed sufficiently to enable them to be categorized as to action potential value.

According to Forbes' early work, sign attention value is also a function of the sign's location or placement. Nowadays, the implementation of standard recommended sign placements has led to experienced road users forming strong expectations as to where signs are likely to be placed. As a consequence, detection is much more likely and more rapid when placement is consistent with expectations than when it is not (Theeuwes, 2002). In effect, particular scanning patterns that coincide with standard sign placement have been selectively reinforced while others have not. Habitual ways of responding to *patterns* of stimuli can also determine priority of attention. Thus, for example, when a number of otherwise equivalent sign elements are present, native English language readers tend to attend to them in the same pattern as reading text, processing the uppermost and left-most elements first (Forbes, 1972).

However, whatever the motivation, needs, expectations, and habits, some stimuli are simply more compelling than others. Relative intensity is one relevant dimension here that was visited when discussing the effects of relative brightness (technically, relative *luminance*: the amount of light entering the eye reflected from a surface; brightness is the subjective response to this and is proportional to it). At night, the use of illuminated signs or retroreflective materials is, of course, useful to enhance sign detection; early research revealed that a reflective intensity of at least 1 mcd/lx cm^2 yielded a 90th-percentile detection distance at 400 m with dipped headlights on an unlit roadway. The effects of dipped headlights from an oncoming car decreased detection distance by up to 120 m (see, for example, Dahlstedt and Svenson, 1976).

Independent of daylight/nightime conditions, however, dynamic variation in stimulus intensity is even more compelling. Sense organs in general and vision and auditory modalities in particular are primed to respond to change. They have been likened to news reporters for whom a steady-state situation is of little interest, but when something changes, they begin to sit up and take notice. Thus, crucial hazard signs often use flashing (as opposed to static) lights (of course, use of flashing lights and sirens varying in pitch and intensity is standard on emergency and police vehicles). Summala and Heitamaki (1984) provide good empirical evidence of the relative power of the addition of a flashing light to a standard road sign in changing driver behavior.

Even without dynamic change in the stimulus, however, sign design may exploit the general principle here by exaggerating intensity variation within the sign. Compare, for example, the attention-getting potential of the two directional signs in Figure 9.14. The lower one attracts more attention because as the eye scans across it, it is stimulated by a high degree of variation in brightness intensity.

It is not always the case, however, that the most conspicuous elements will attract attention. Laboratory studies by Theeuwes (1991) and field research by Hughes and Cole (1986) demonstrate that drivers do not necessarily look at conspicuous objects that are irrelevant to the task in hand (see discussion by Theeuwes, 2002). As a consequence, bright lights in an urban environment do not necessarily distract the driver from driving-relevant stimuli. Nevertheless, recent laboratory studies by Ho et al. (2001) indicate that scenes rated by observers as high in clutter require longer latencies and more fixations to acquire a target sign, are associated with more errors, and have longer fixation durations than scenes rated as low in clutter. This is particularly the case with older drivers. Although they argue that this research should be replicated in field studies, Ho et al. recommend that engineers consider reducing

FIGURE 9.14 Dynamic stimulation in a static display.

high clutter by avoiding redundant signs and competing advertisement signs and enhancing the relative conspicuity of safety-relevant signs.

As discussed by several authors in this book, one promising piece of assistive technology that could help solve the problem of failure to detect a sign is the automatic capture, hold, and display of safety-critical sign information in the driver's vehicle (Fuller and Santos, 2002). Recent work in Sweden has evaluated a GPS-based system for transmitting to a driver the speed limit pertaining to the roadway section in which he is traveling. Early results indicate that this induces greater compliance with the limit, avoiding situations in which the driver unintentionally exceeds it. Furthermore, drivers appear to accept the additional information (Gregersen, 2003).

Such a solution could be further enhanced by the use of associated automatic auditory warnings. These could provide a general alert, such as the "attenson" used in aircraft cockpits (Hawkins, 1987), or give brief specific messages to the driver. At particularly hazardous locations, if approach speed is too high, the message might tell the driver the critical feature of the roadway and an appropriate speed, rather in the manner of the navigating co-driver in a rally car. Auditory warnings, of course, do not depend on where the driver is looking in order for them to be noticed and may be used, for example, to call attention to the need to examine a displayed visual warning. Research suggests that the presence of a voice warning can produce strong and reliable increases in compliance (Wogalter et al., 2002).

9.3 CONCLUSIONS AND PRACTICAL IMPLICATIONS

This chapter has focused on the vital issue of motivation with respect to traffic signs. As discriminative stimuli, signs can provide the driver with advance information of required behavior in order to sustain a safe outcome. However, as potential controlling elements of driver behavior, signs are, to varying extents, in competition with naturally occurring discriminative stimuli; they frequently signal uncertain contingencies between behavior and its consequences. To be more effective, signs need to represent in a reliable way prevailing contingencies. However, when compliance with posted speed restriction is required, this may mean introducing additional contingencies, such as enforced penalties, to those occurring in the natural environment. To be effective, signs must, of course, also be detected and recognized. Although research has generated enough knowledge in this area to enable society to be prescriptive about detailed features of sign design and location, the exploitation of new assistive technologies has the potential to enhance significantly the impact of traffic signs on driver behavior.

REFERENCES

Al-Madani, H. and Al-Janahi, A-R. (2002). Assessment of drivers' comprehension of traffic signs based on their traffic, personal and social characteristics, *Transp. Res. Part F,* 5, 63–76.

Crundall, D. and Underwood, G. (2002). The priming function of road signs, *Transp. Res. Part F*, 4, 187–200.

Dahlstedt, S. and Svenson, O. (1976). Detection distances of retroreflective road signs during night driving, *Goteborg Psychol. Rep.*, 4, 6, 4.1–4.9.

Forbes, T.W. (1972). Visibility and legibility of highway signs, in T.W. Forbes (Ed.), *Human Factors in Highway Traffic Safety Research*. New York: Wiley, 95–109.

Forbes, T.W. and Holmes, R.S. (1939). Legibility distances of highway destination signs in relation to letter height, letter width and reflectorization, *Proc. Highway Res. Bd.*, 19, 321–335.

Fuller, R. and Santos, J.A. (2002). *Human Factors for Highway Engineers*. New York: Pergamon Press.

Gardner, D.J. and Rockwell, T.H. (1983). Two views of motorist behavior in rural freeway construction and maintenance zones: the driver and the state highway patrolman, *Hum. Factors*, 25, 415–424.

Gregersen, N.P. (2003). Personal communication.

Groeger, J.A. (2000). *Understanding Driving: Applying Cognitive Psychology to a Complex Everyday Task*. Hove: Psychology Press.

Hagemeister, C. and Westhoff, K. (1997). Reinforcing safer car driving, in T. Rothengatter and E. Carbonnell Vaya (Eds.), *Traffic and Transport Psychology: Theory and Application*. Oxford: Pergamon Press, 403–411.

Hawkins, F.M. (1987). *Human Factors in Flight*, Aldershot: Gower.

Helliar–Symons, R.D. (1983). Automatic close-following warning sign at Ascot, TRRL Laboratory Report 1095, Crowthorne: DoE, DoT.

Ho, G., Scialfa, C.T., Caird, J.K., and Graw, T. (2001). Visual search for traffic signs: the effects of clutter, luminance and aging, *Hum. Factors*, 43, 2, 194–207.

Holland, C.A. and Conner, M.T. (1996). Exceeding the speed limit: an evaluation of the effectiveness of a police intervention, *Accident Anal. Prev.*, 28, 5, 587–597.

Howarth, C.I. (1988). The relationship between objective risk, subjective risk and behavior, *Ergonomics*, 31, 527–535.

Hughes, P.K. and Cole, B.L. (1986). What attracts attention when driving? *Ergonomics*, 29(3), 377–391.

Johansson, G. and Backlund, F. (1970). Drivers and road signs, *Ergonomics*, 13, 749–759.

Johansson, G. and Rumar, K. (1966). Drivers and road signs: a preliminary investigation of the capacity of car drivers to get information from road signs, *Ergonomics*, 9, 57–62.

Lajunen, T., Hakkarainen, P., and Summala, H. (1996). The ergonomics of road signs: explicit and embedded speed limits, *Ergonomics*, 39(8), 1069–1083.

Macdonald, W.A. and Hoffmann, E.R. (1991). Drivers' awareness of traffic sign information, *Ergonomics*, 34, 5, 585–612.

Milosevic, S. and Gajic, R. (1986). Presentation factors and driver characteristics affecting road-sign registration, *Ergonomics*, 29, 6, 807–815.

Schieber, F. and Goodspeed, C.H. (1997). Nightime conspicuity of highway signs as a function of sign brightness, background complexity and age of observer, in *Proc. Hum. Factors Ergonomics Soc. 41st Annu. Meet.*, Santa Monica, CA: Human Factors and Ergonomics Society, 1362–1366.

Shinar, D. and Drory, A. (1983). Sign registration in day time and night time driving, *Hum. Factors*, 25, 117–122.

Stuss, D.T., Shallice, T., Alexander, M.P., and Picton, T.W. (1995). A multi-disciplinary approach to anterior attentional functions, in J. Grafman, K. Holyoak, and F. Boller (Eds.), *Structure and Function of the Human Prefrontal Cortex*, *Ann. N.Y. Acad. Sci.*, 279, 191–211.

Summala, H. and Hietamaki, J. (1984). Drivers immediate responses to traffic signs, *Ergonomics*, 27, 2, 205–216.

Summala, H. and Naatanen, R. (1974). Perception of highway traffic signs and motivation, *J. Safety Res.*, 6, 4, 150–154.

Theeuwes, J. (1991). Visual selection: exogenous and endogenous control, in A. Gale, I. Brown, C. Haslegrave, H. Kruysse, and S. Taylor (Eds.), *Vision in Vehicles III*. Amsterdam: North Holland, 53–62.

Theeuwes, J. (2002). Sampling information from the road environment, in R. Fuller and J.A. Santos (Eds.), *Human Factors for Highway Engineers*. Amsterdam: Pergamon Press, 131–146.

U.K. Driving Standards Agency (1992). *Driving Skills: the Driving Manual*, London: HMSO.

Wilde, G.J.S. (1994). *Target Risk*, Toronto: PDE Publications.

Wogalter, M.S., Conzola, V.C., and Smith–Jackson, T.L. (2002). Research-based guidelines for warning design and evaluation, *Appl. Ergonomics*, 33, 219–23.

COLOR FIGURE 1.2a Warning signs in Australia and Spain in the traffic environment.

Anglo-Saxon and Spanish version
of obligatory direction sign

Anglo-Saxon and Spanish version
of double bend sign

COLOR FIGURE 1.2b Examples of the lack of harmonization in traffic sign designs around the world.

COLOR FIGURE 4.4 A possible example of the dominant–subordinate large–small delay.

(a)

(b)

COLOR FIGURE 4.6 An example of a cluttered signing environment in Spain.

COLOR FIGURE 6.1 Basic lower-quadrant mechanical semaphore signal showing the STOP and PROCEED aspects, respectively.

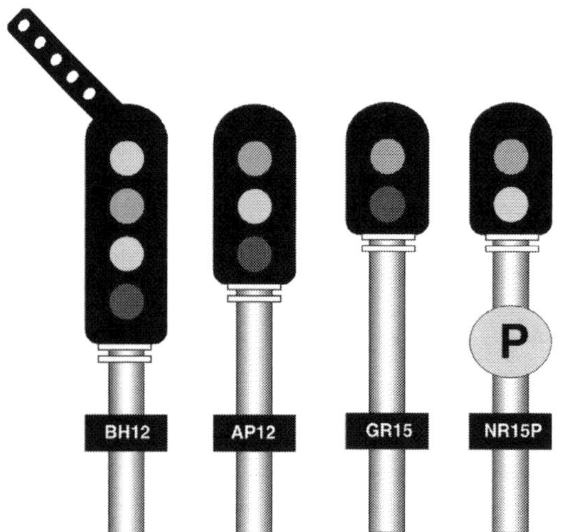

COLOR FIGURE 6.2 Examples of basic colored light signals (following QR route-signaling standards based on British standards). From left to right: four-aspect signal with left turn-out junction route indicator attached; three-aspect signal; two-aspect signal; and two-aspect signal repeater.

COLOR FIGURE 6.9 Dynamic speed indicator (DSI) on a four-aspect route signal. In this example, the DSI is lit, showing a 40-km/h speed limit. The DSI is used only if the junction route indicator (JRI) "rabbit-ear" is lit. This allows a train to proceed through a turnout at 40 km/h when, normally, the speed may have been assumed to be 25 km/h.

COLOR FIGURE 6.11 QR passenger information display system — an example of an online real-time information system.

COLOR FIGURE 7.6 Taxiway signs on location. (Courtesy of Bo Wiberg, Swedflight Design Group' www.swedflight.com.)

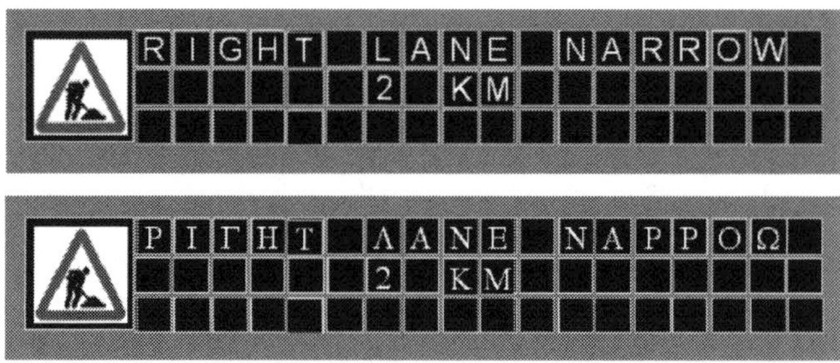

Causes and consequences

Consequences and causes

COLOR FIGURE 12.1 Considering consequences vs. causes for national vs. international drivers.

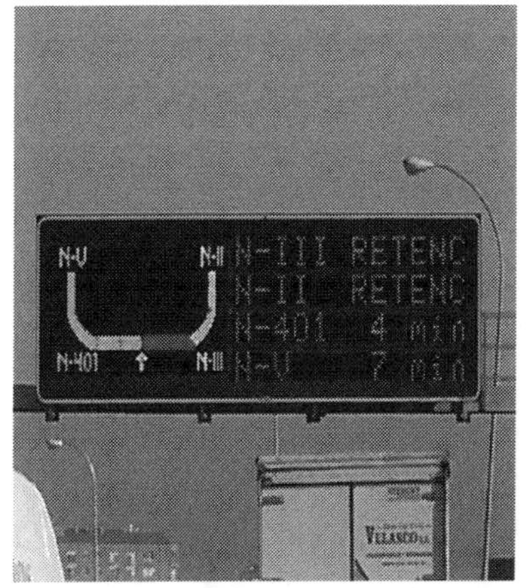

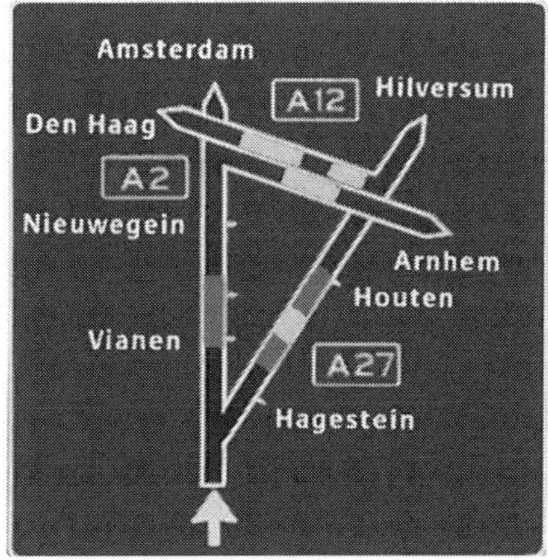

COLOR FIGURE 13.1 Real-time graphic displays showing information on congestion and travel time. Left: traffic mimicry panels in M-40 (Madrid); right: real-time graphic display (Dutch prototype).

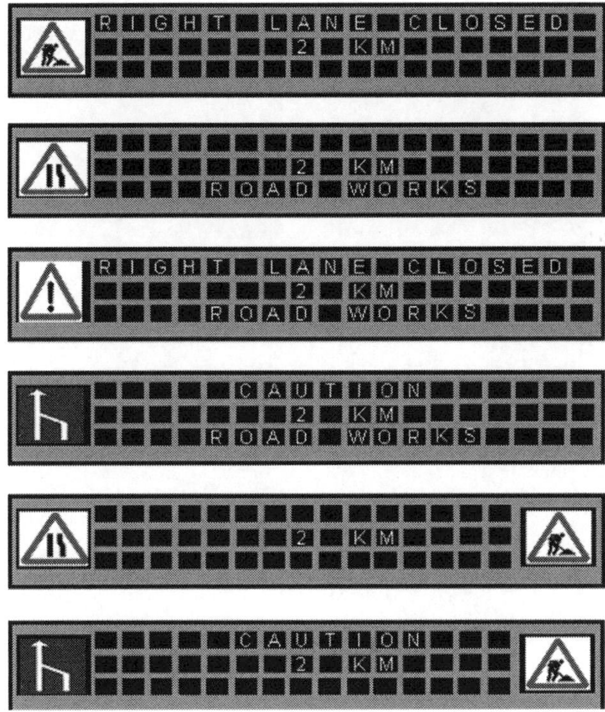

COLOR FIGURE 13.2 Examples of different 'correct' VMS alternatives for indicating "Caution. The right lane is closed in 2 km due to road work." Pictograms are in agreement with 1968 Vienna Convention; formats are in agreement with WERD/DERD (2000).

Text only VMS

Pictogram VMS

Combined pictogram-text VMS

Combined pictogram-text VMS

Combined (vertical layout)

Combined (horizontal layout)

COLOR FIGURE 13.3 Examples of VMS with different layout, structure for informational elements, and disposition on the panel.

10 Cross-Cultural Uniformity and Differences in Roadway Signs, Evaluation Techniques, and Liabilities

Hashim Al-Madani

CONTENTS

10.1 Introduction 156
10.2 How Satisfactory Is the Existing Signing System? 156
10.3 Techniques Used in Evaluating and Measuring Sign Comprehension 157
10.4 Uniformity and Differences in International Signing Systems 161
10.5 Legislation and Liabilities 164
10.6 Conclusions and Implications 165
References 165

Posted traffic signs convey messages in words and/or symbols and are erected to regulate, warn, or guide road users. Traffic signs, however, are most effective when they fulfill a need, command attention, convey a clear and simple meaning, command respect of road users, and give adequate time for proper response (Federal Highway Administration MUTCD, 2001; Canfield, 1999). Questions such as how well the existing sign systems are performing in terms of comprehension are discussed in this chapter. Techniques used in evaluating and measuring drivers' comprehension of signs, standardization uniformity and differences among countries, and legislation and liabilities are also covered.

0-415-31086-5/04/$0.00+$1.50

10.1 INTRODUCTION

Traffic signs have been a topic of considerable interest to researchers in many parts of the world. Researchers cover a wide range of aspects related to engineering, traffic safety, education, and human physical and psychological capabilities. One of the earliest publications on posted signs is a manual and specification for the manufacture, display, and erection of road markings and signs published by the U.S. Ministry of Transport in 1927. As seen earlier in this book, more literature is also available in Europe and America. In fact, a wide range of techniques has been used to evaluate roadway posted signs.

In the 1940s and early 1950s, much of the research in this field consisted of technical studies, such as those related to impact assessment, cost estimation, signs' placement, accident occurrences, and evaluation and illumination effectiveness (e.g., Riegelneier, 1942; Lauer and McMonagle, 1955). In the early 1960s researchers started to give more weight to human-related studies, particularly those related to motorists' behavior (Hakkinen, 1965; Johansson and Rumar, 1966). However, studies on drivers' comprehension of traffic signs from psychological and demographic points of view are still scarce, especially when cross-cultural differences are considered.

10.2 HOW SATISFACTORY IS THE EXISTING SIGNING SYSTEM?

Road signs as a communication tool for drivers' navigational purposes are not fully satisfying drivers' needs yet, especially from the point of view of their comprehension. Many researchers since the 1960s have reported that the road sign system does not fulfill its intended and assumed function in a satisfactory way (Johansson and Rumar, 1966; Womack et al., 1981; Summala and Hietamäki, 1984; Ogden et al., 1990; Fisher, 1992; Al-Madani and Abdul-Ghani, 1995).

Summala and Hietamäki (1984) and Summala and Näätänen (1974) attributed the defective function of the traffic sign system mainly to motivational factors. Drivers generally have perceptual skills sufficient to detect signs but they do not feel the need for such information (Al-Madani and Al-Janahi, 2002b). Al-Madani and Al-Janahi (2001, 2002a) clearly noticed serious problems in understanding the existing sign system by drivers from Western, Asian, and Arab countries. Knoblauch and Pietrucha (1987) also identified deficiencies in understanding symbols that could pose safety and operational problems. In America, Sanders et al. (1973) found deficiencies in motorists' comprehension of some of the common traffic control devices, including signs. Al-Yousifi (1999, 2002a, b) indicated several problem areas in existing signs, among which ambiguous signs, symbols with different meanings, and symbols confusing to motorists were the major ones.

Johanson and Backlund (1970) found that sign recognition varies widely among drivers. This phenomenon has been clearly observed in many other recent studies (Hawkins et al., 1993; Al-Madani and Abdul-Ghani, 1995; Al-Yousifi, 1999; Shinar et al., 1999). Jabbar and Naqvi (1992) found that drivers make many more significant errors in detecting symbolic signs than with alphanumeric ones. Alphanumeric signs are better, in terms of drivers' reaction time, when compared to symbolic ones, as are warning signs compared with regulatory signs (Dewar et al., 1976). Ogden et al. (1990) revealed that motorists have some difficulty interpreting both word and symbol messages on signs. Al-Madani and Al-Janahi (2002a, b) observed the existence of functional problems in Arab, Asian, European, and American drivers' understanding of regulatory signs compared with warning signs.

Although signs adopted in the Arab region are similar to those used worldwide and comply with the Vienna Convention on road signs and signals (United Nations Economic Commission for Europe, 1968, 1974, 1995), drivers fail to comprehend them properly (Al–Madani and Al-Janahi, 2002a). Avant et al. (1986) observed unconscious errors in recognition of signs. Johansson and Rumar (1966) and Fisher (1992) raised serious questions concerning the applicability of the existing signing system for current conditions. Therefore, it is strongly recommended to evaluate drivers' understanding of the symbols used on roadway signs systematically and comprehensively so as to know where best to concentrate efforts (Dewar et al., 1994). For example, little is known about the effectiveness of signs for older drivers; Kline and Dewar consider this topic in a chapter in this book.

10.3 TECHNIQUES USED IN EVALUATING AND MEASURING SIGN COMPREHENSION

As mentioned earlier in this book, the vast range of techniques used in evaluating traffic signs makes comparison of the results difficult because of differences in methods, samples, limitations, definitions, and assumptions. Table 10.1 summarizes some of the studies analyzing traffic signs carried out by various researchers.

In any self-reported technique, biases against those who did not respond may occur in questionnaire-based studies because they may ignore certain categories of society who did not return the questionnaires. Furthermore, Dewar (1994) expressed concern about multiple-choice tests in which the respondents are faced with several choices for selection, one of which is correct; the quality of the distracters (wrong answers) could greatly influence comprehension scores. Answers that could easily be ruled out enable the respondents to choose the correct answers more easily. This would unfairly inflate the comprehension level compared with open-ended tests (Hawkins et al., 1993; Dewar, 1994) or with a test including more plausible distracters.

Wolff and Wogalter (1998) compared responses to questionnaires that employed all three methods. They found that the multiple-choice test with less plausible distracters inflated comprehension scores by 30% compared with multiple-choice tests with more plausible distracters or open-ended tests, which did not differ. Furthermore, multiple-choice methods eliminate drivers' freedom to develop their own explanations of a device because their responses are influenced by the given choices, thus eliminating potential areas of confusion (Hawkins et al., 1993). However, this method reduces answering time and analysis cost. The method assists respondents who lack writing and expressing abilities but act correctly toward the

TABLE 10.1
Techniques Used by Various Researchers to Evaluate Traffic Signs, with Sample Sizes and Covered Signs

No.	Author	Year	Country	Sample Size	Signs Covered	Technique Used	Testing Method	Survey Site
1	Lauer and McMonagle	1955	U.S.	170 drivers	Not specified	Simple landscape of road-side features	Observing experimental drivers	Laboratory
2	Johansson and Rumar	1966	Sweden	1031 drivers (randomly selected)	4 warning, 1 regulatory	In-car test	Pushing a button as information is gained	170-km stretch of road
3	Ells and Dewar	1979	U.S. (Calgary)	6 male and 6 female drivers (students)	8 regulatory, 8 warning and verbal messages	Projected slides	Identification speed using yes or no	University campus
4	Hulbert and Fowler	1979	Across U.S.	3100 drivers	8 signs, signals and 8 markings	Video depicting real situation from a vehicle	—	Not mentioned
5	Hulbert et al.	1980	Across U.S.	—	10 signs and 8 markings	Video depicting real situation from a vehicle	—	Not mentioned
6	Allen et al.	1980	U.S.	—	72 signs	Dynamic highway scene from a driving simulator	—	Laboratory
7	Halpern	1984	U.S. (California)	20 male and 20 female drivers	8 symbol signs and 7 verbal messages	Projected slides	Observing experimental drivers for verbal discrimination of signs	University campus and surroundings
8	Knoblauch and Pietrucha	1987	U.S.	Variety of professionals related to roads	30 signs	Comments on existing signs	Open discussion	Offices and laboratory
9	Richards and Heathington	1988	U.S. (Tennessee)	176 drivers, 35 police officers (random)	Railroad grade crossing traffic control devices (signs and markings)	Pictorial	16 choice-based questions (multiple choice) and 1 short-answer one	(License station and university)

10	Ogden et al.	1990	USA (Texas–Houston)	205 drivers from farm-to-work area	Work-zone signs	Videotape of 35-mm photographs of signs	Choice-based questions and personal interviews	Licensing stations and shopping malls
11	Fisher	1992	S. Africa (Johannesburg)	96 drivers (random)	Various signs in car on the highway	17 minutes videotape from 35-mm slides	Recall of passed signs (question the driver)	9 km across country
	Hawkins et al.	1993	U.S.	1745 drivers	13 regulatory, 18 warning, 7 pavement markings, 8 other devices		Choice-based (multiple-choice) questions	Licensing stations
12	Dewar Dewar et al.	1994 1994	U.S. (Texas and Idaho), Canada (Alberta).	I. 480 drivers II.219 drivers (volunteers)	11 regulatory signs, 38 warning signs, and 36 other devices.	Projected color slides	Open-ended questions on the meaning of the signs.	Licensing station, service area, & recreational areas.
13	Al-Madani and Abdul-Ghani	1995	Bahrain	I.950 drivers	10 regulatory,	Questionnaire with colored pictures	Choice-based (multiple choice) on correct sign & short-answer questions on drivers' characteristics.	Random distribution to place of work.
	Al-Madani Al-Madani and Al-Janahi	2000 2002	5 Arab countries (Bahrain, Kuwait, Qatar, U.A.E., Oman).	II.4774 drivers (stratified random sampling)	18 warning, 3 fake signs			
14	Al-Yousifi	1999, 2002	France, Kuwait, Bahrain & Iran	4000 drivers	More than 17 regulatory and warning signs	Questionnaire with pictures	Choice-based (multiple-choice) questions	Not specified

conveyed messages in real life. It also eliminates judgmental errors in answers when different judges code the responses because the distinction between wrong and correct answers is clear (Al-Madani and Al-Janahi, 2002b).

These devices serve little purpose if they are not understood well by drivers. In fact, the American National Standard Institute (ANSI, 1991) and the International Organization for Standardization (ISO, 1984) advise that symbols must meet a criterion of at least 85 or 67% correct, respectively, in a comprehension test to be considered acceptable (Wolff and Wogalter, 1998). Understanding of signs in general varies considerably among drivers (Johansson and Backlund, 1970). The level of drivers' comprehension of individual signs, like seat belt usage, differs from one country to another; it depends on many factors, such as the licensing system, level of motorization, literacy, and level of education. Hulbert and Fowler (1979) found respondents' level of comprehension in the U.S. to be about 75%. In Texas, Idaho, and Alberta (Canada), 95% of drivers understood fewer than 20% of the posted signs. Furthermore, about 40% of these drivers understood as few as 10% of the signs (Hulbert et al., 1980).

Dewar et al. (1994) found that American and Canadian drivers comprehended 81, 75, and 59% of regulatory, warning, and school signs, respectively. American and European drivers in the Arab world identified 78 and 69% of regulatory and warning signs (Al-Madani and Al-Janahi, 2001; Al-Madani, 2001). Although these findings seem to match, Hawkins et al. (1993) found American drivers' comprehension of signs to be much lower than the figures mentioned earlier. Drivers' correct response to regulatory signs, warning signs, and pavement markings were 64, 61, and 67%, respectively.

In the Arab world drivers comprehended just over half of the signs posted along roadways. Knowledge of regulatory and warning signs was 55 and 56% (Al-Madani and Al-Janahi, 2002a). The true average, in the U.S. and Arab worlds would have been even lower if the respondents had not been supported with selection choices. The signs best comprehended by drivers in the Arab world regardless of their exposure rates were found to be those indicating (Table 10.2) no right turn (identified by 92% of the drivers), dual carriage with three lanes (the right hand lane is closed), no U-turn, and slippery road. Similarly, signs least understood were as follows: no pedestrians are allowed to pass, diversion to opposite, staggered junctions, and vehicles may pass on either way (identified by 16% of drivers). The tested signs included two regulatory signs indicating "end of prohibition" order. Both proved to be among the least-understood signs by drivers (Al-Madani and Al-Janahi, 2001).

Results for America and Canada are summarized in Table 10.3; Table 10.4 summarizes drivers' comprehension of signs in the Arab world based on their socioeconomic characteristics (Al-Madani, 2001). As a note of caution suggested by Fisher (1992), failure to recall or understand roadway signs does not guarantee a failure to act properly in response to the message. On the other hand, incorrect responses to signs should not be fully equated to failure of the traffic control device (Hawkins et al., 1993).

TABLE 10.2
Drivers' Comprehension of Individual Regulatory Signs in the Arab World

Sign Title	Drivers' Correct Identification of Sign (%)	Drivers' Correct Identification of Sign (%/1000 km)
Axle weight limit	38.3	25.5
Mini roundabout	35.0	0.8
No pedestrians are allowed to pass	30.3	3.4
Turn left	73.6	0.5
Maximum height limit	74.1	0.8
Maximum speed limit	77.3	0.1
No stopping	36.6	0.7
No right turn	92.2	1.8
No waiting	41.0	3.7
End of prohibition of overtaking	47.5	17.6
No U-turn	85.9	7.8
Priority to oncoming traffic	47.6	23.8
Closed to motor cars	62.5	62.5
End of prohibition of goods vehicles	32.6	6.5
No entry for vehicular traffic	72.0	0.7
Turn left ahead	77.9	0.6
Vehicles may pass either side	16.3	1.6
Keep right	62.1	0.4
Staggered junction	21.4	14.2
Traffic merges from left	38.7	0.3
Road narrows from both sides	76.5	51.0
Dual carriage with three lanes — right hand lane is closed	86.0	6.6
Slippery road	84.2	84.2
Dual carriageway ends	67.4	44.9
Diversion to opposite	29.1	29.1
Road narrows on left ahead	80.2	4.0
Hump bridge ahead	38.7	38.7
Other danger plate indicates nature of danger	41.5	0.4

Source: Adapted from Al-Madani and Al-Janahi 2001 .

10.4 UNIFORMITY AND DIFFERENCES IN INTERNATIONAL SIGNING SYSTEMS

Ministers from 18 European countries met in Vienna in 1968 under the umbrella of the United Nations to agree on a uniform sign standard for worldwide applications. They ended up with a convention now known as the Vienna Convention on Road Signs and Signals (United Nations Economic Commission for Europe, 1968), which is widely accepted by traffic authorities and employed by most countries in the

TABLE 10.3
Drivers' Comprehension of Individual Signs in U.S. and Canada

U.S. and Canada		U.S. (Texas)	
Sign Title	**Comprehension (%)**	**Sign Title**	**Comprehension (%)**
No right turn	96.9	Yield	79.4
No U-turn	97.9	Speed zone	55.0
Straight or left	16.0	Double turn	65.0
No parking	95.4	Two-way left turn	58.6
Keep right	85.8	HOV restriction	45.7
No trucks	97.1	Keep right	69.9
Right turn	91.7	Stop ahead	87.4
Right curve	94.9	Reverse turn	66.5
Right reverse turn	70.4	Turn right	31.9
Yield ahead	75.8	Left curve	32.4
Stop ahead	90.2	Lane ends merge left	64.0
Added lane	25.5	Narrow bridge	81.7
Merge	90.0	Lane reduction transition	61.2
Lane reduction transition	38.1	Divided highways ends	50.7
Narrow bridge	77.3	Limited sight distance	44.9
Divided highway	78.3	Slippery, when wet	62.3
Divided highway ends	71.7	No passing zone	88.0
Two-way traffic	89.2	Double solid white line	61.0
Slippery (when wet)	44.6	Solid white edge line	74.7
Double arrow	34.8	Single broken yellow center line	76.8
Pedestrian crossing	91.9		
Low vertical clearance	96.6		
Side road (right, 45°)	72.5		
Side road (right, 90°)	87.4		

Sources: U.S. and Canada: adapted from Dewar, R.E. et al., *Transp. Res. Rec.*, 1456, 1–10, 1994; U.S. (Texas): adapted from Hawkins, H.G. et al., *Transp. Res. Rec.*, 1403, 67–82, 1993.

world. Several amendments have also followed the original convention. These accommodate new signs and necessary improvements; see the latter chapters in this book discussing Variable Message Signs. The U.S. follows its own manual for traffic signs, *Manual on Uniform Traffic Control Devices for Streets and Highways* (MUTCD; Federal Highway Administration, 2001), for the use of federal government and other related agencies.

Although the MUTCD in America is continuously evaluated and updated at least once each decade, the Vienna Convention has remained without substantial changes since 1973 (Al-Yousifi, 1999, 2002a,b). Although some traffic signs are uniform all over the world, many countries have issued their own sign manuals (Al-Madani, 2000). In addition to the Vienna Convention, the MUTCD (FHWA, 2001), and the European Rules Concerning Road and Traffic Signs and Signals (United Nations Economic Commission for Europe, 1974), many other countries have produced their

TABLE 10.4
Drivers' Perception of Regulatory and Warning Signs by Their Socioeconomic Characteristics

Parameters	Categories	Perception of Traffic Signs (%)	
		Regulatory Signs	Warning Signs
Monthly income, tax free — in B.D. (U.S. $)	<300 (<800)	53.0	53.9
	300–500 (800–1320)	55.0	55.4
	501–700 (1321–1850)	55.5	56.6
	>700 (>1850)	57.9	61.4
Age (years)	16–24	50.1	50.8
	25–34	54.2	55.0
	35–44	58.5	60.1
	45–55	59.8	61.9
	55+	55.6	49.0
Years of education (education level)	<1 (Uneducated)	46.9	49.4
	1–11 (Below secondary)	50.0	50.1
	12–13 (Secondary)	53.3	54.3
	14–15 (Diploma)	54.4	54.6
	16–18 (B.S.)	57.3	58.5
	18+ (Higher degree)	61.7	65.6
Gender	Male	56.6	57.5
	Female	49.7	51.5
Marital Status	Single	52.3	52.1
	Married	56.5	58.1
Nationality	GCC	51.7	52.1
	Arab	56.7	55.8
	Asian	61.5	65.2
	EC + U.S.	69.1	78.3
	African and Oceanic	62.5	68.7

Source: Adapted from Al-Madani, H.M.N., *Perceptual Motor Skills*, 92, 72–82, 2001.

own manuals and specifications. In Australia, the Australian Standard Rules for the Design, Location, Erection, and Use of Road Signs and Signals (Standard Association of Australia, 1960) is employed. In Canada, the Manual on Uniform Traffic Control Devices for Canada (Road and Transportation Association of Canada, 1966) is used. Great similarities do exist among all these standards; however, effort should be continued to promote international conformity (Pline, 1992).

In America and Canada, most warning signs are shaped like a diamond; in most other countries the shape is triangular in conformity with the Vienna Convention (United Nations Economic Commission for Europe, 1968). The former provides more space for any future research needs (Al-Yousifi, 2002b). Furthermore, an oblique bar reinforces the regulatory signs indicating prohibitory and restrictive actions according to the MUTCD but not in most of the signs in accordance with Vienna Convention or European rules. Although such signs are infrequent in France, Iran, Kuwait, and Bahrain, drivers in these countries prefer the inclusion of the oblique bar on warning signs because it enhances recognition of the prohibition action (Al-Yousifi, 2002a). In order to bridge some of these differences in the signs' shapes between those indicated in the MUTCD (mainly used by Americans, Canadians, and some other countries) and those that appeared in the original Vienna Convention (utilized by Europeans and many other countries), an Amendment to the Vienna Convention (United Nations Economic Commission for Europe, 1995) was introduced to treat such differences by allowing both to be acceptable.

To bridge the gap in comprehension of signs between drivers from different countries (drivers in America and Europe comprehend posted signs better than drivers from most of the other countries) (Al-Madani and Al-Janahi, 2001) and to improve their signing skills, manufacturers, designers, and researchers in the field of traffic signs (mostly from developed countries) should reconsider the design of many existing signs. This is particularly necessary for signs that are poorly comprehended by drivers of other nations. An improved signing system that suits the need of drivers in less motorized countries is certainly important (Al-Madani, 2001; Al-Yousifi, 1999, 2002a). The improved signing system should convey regulatory and warning information correctly to drivers in developing and less developed countries. On the other hand, authorities in the latter countries should also improve their traffic education system in order to raise drivers' comprehension skills for traffic signs.

10.5 LEGISLATION AND LIABILITIES

Public and private transportation organizations as well as individual employees of these organizations are increasingly exposed to the possibility of litigation. This can range from being a witness to being a defendant in a civil liability lawsuit (Van Gelder et al., 1992). In the U.S., although the federal government is the prime developer of the MUTCD (Federal Highway Administration, 2001), it is held harmless by the courts for injuries and losses due to infringements of standards, changes in design, and so on. In fact, MUTCD is a standard for the installation, design, and application of control devices; it does not hold any legal requirements for installations.

The use of such devices is based on an engineering judgment for problematic locations (Burnham, 1992). Lawsuits against the government for negligence of design of signs and operation generally fail (Van Gelder et al., 1992). However, failure to maintain safe conditions for users that results in injury or property loss will subject the agency with prime responsibility of posting signs to liability (Burnham, 1992; Wright, 1996). In the area of design of signs, an agency is generally held immune from tort liability as long as discretionary authority in the development of the design has not been abused. In the areas of sign implementation, operations, and maintenance, any breaches of duty, particularly negligence-related ones, expose the relevant agency to liability (Van Gelder et al., 1992). Road users should comply with the instructions conveyed by road signs and markings, even if these instructions appear to contradict other traffic regulations (United Nations Economic Commission for Europe, 1974).

Agencies responsible for traffic signs and markings can reduce their liability risks in many ways. For example, they should comply with federal and state governments' standards and specifications in design, placement, and installation of such devices (Burnham, 1992). They should replace them as their expected life ends, keep an inventory of devices (Wright, 1996), and review and improve them continuously.

10.6 CONCLUSIONS AND IMPLICATIONS

Authorities responsible for roadway facilities erect various traffic signs to convey the necessary messages in words and/or symbols to regulate, warn, or guide road users. These tools serve little function if they are not comprehended well by road users. Although many individual signs are well comprehended by drivers in various parts of the world, many others are not. Serious questions are raised concerning the applicability of many of the existing signs for current conditions. Although it would be difficult for all drivers to achieve a complete understanding of all traffic signs, there is a large potential for significant improvement. Special attention should be paid to improving the means of learning signing systems when driving licenses are obtained. In the area of design of signs, an agency is generally held immune from tort liability as long as discretionary authority in the development of the design has not been abused.

The continued application of the existing signing system, which is a one-way communication system between the road user (who is the active member) and the sign, is questionable. Future needs call for at least two-way systems; this point is taken up by several chapters in this book.

REFERENCES

Allen, R.W., Parseghain, Z., and van Valkenburg, P.C.A. (1980). Simulator evaluation of age effects on symbol sign recognition. *Proc. Hum. Factor Soc., 24th Annu. Meet.* Los Angeles: Human Factors Society.

Al-Madani, H.M.N. (2000). Influence of drivers' comprehension of posted signs on their safety related characteristics. *Accident Anal. Prev.*, 32, 575–581.

Al-Madani, H.M.N. (2001). Prediction of drivers' recognition of posted signs in five Arab countries. *Perceptual Motor Skills*, 92, 72–82.

Al-Madani, H.M.N. and Abdul-Ghani, A. (1995). Characteristics of drivers' understanding of posted signs. *Proc. Int. Forum Road Res.*, Bangkok, Thailand, Swedish National Road and Transport Research Institute, Sweden, 193–202.

Al-Madani, H.M.N. and Al-Janahi, A.R. (2001). Differences in traffic signs' recognition between drivers of different nations. *Proc. Traffic Saf. Three Continents, Moscow, Swedish National Road, and Transport Research Institute*, Conference 18A, Sweden, 5, 404–424.

Al-Madani, H.M.N. and Al-Janahi, A.R. (2002a). Assessment of drivers' comprehension of traffic signs based on their traffic, personal and social characteristics. *Transp. Res. F*, 5, 63–76.

Al-Madani, H.M.N. and Al-Janahi, A.R. (2002b). Role of drivers' personal characteristics in understanding traffic sign symbols. *Accident Anal. Prev.*, 34, 185–196.

Al-Yousifi, A.E. (1999). Investigation of traffic signs to improve road safety. *Proc., 10th Int. Conf. Traffic Saf. Two Continents.* Malmo, Sweden, 773–787.

Al-Yousifi, A.E. (2002a). A closer look at our traffic signs. Safety on Roads, 2nd International Conference, SORIC '02, Center for Transport and Road Studies, University of Bahrain, Paper no. E147, Bahrain.

Al-Yousifi, A.E. (2002b). *An Examination of Perceptual Patterns for Traffic Signs Recognition in Four Countries: an Empirical Approach.* College of Business, American University in London, U.K.

American National Standards Institute (ANSI) (1991). Accredited standards of safety colors, signs, symbols, labels and tags, Z535.1. Washington D.C: National Electric Manufacturers Association.

Avant, L.L., Brewer, K.A., Thieman, A.A., and Woodman, W.A. (1986). Recognition errors among highway signs. *Transp. Res. Rec.*, 1027, 42–45.

Burnham, A.C. (1992). Traffic signs and markings. In Pline, J. (Ed.), *Traffic Engineering Handbook*, 4th ed. Washington, D.C.: Institute of Transportation Engineers, 239–258.

Canfield, R.R. (1999). Traffic signs and markings. In Pline, J. (Ed.), *Traffic Engineering Handbook*, 5th ed. Washington, D.C.: Institute of Transportation Engineers, 411–452.

Dewar, R.E. (1994). Design and evaluation of graphic symbols. *Proc. Public Graphics*, Utrecht, Netherlands: University of Utrecht, Department of Psychonomics, 24.1–24.18.

Dewar, R.E., Ells, J.G., and Mundy, G. (1976). Reaction times as an index of traffic sign perception. *Hum. Factors,* 18(4), 381–392.

Dewar, R.E., Kline, D.W., and Swanson, H.A. (1994). Age differences in comprehension of traffic sign symbols. *Transp. Res. Rec.*, 1456, 1–10.

Ells, J.G. and Dewar, R.E. (1979). Rapid comprehension of verbal and symbolic traffic sign message. *Hum. Factors*, 21, 61–168.

Federal Highway Administration (2001). *Manual on Uniform Traffic Control Devices for Streets and Highways* (MUTCD), millennium ed., Department of Transportation, FHWA, Washington D.C. http://mutcd.fhwa.dot.gov. http://mutcd.fhwa.dot.gov/kno-overview.htm. Retrieved January 4, 2003.

Fisher, J. (1992). Testing the effect of road traffic signs information value on driver behavior. *Hum. Factors*, 34(2), 231–237.

Hakkinen, S. (1965). Perception of highway traffic signs. Report 1, Helsinki, Talja, Meddelar, 4, 6–12.

Halpern, D.F. (1984). Age differences in response time to verbal and symbolic traffic signs. *Exp. Aging Res.*, 27(2), 201–204.

Hawkins, H.G., Womak, K.N., and Mounce, J.M. (1993). Driver comprehension of regulatory signs, warning signs and pavement markings. *Transp. Res. Rec.*, 1403, 67–82.

Hulbert, S. and Fowler, P. (1979). Motorists' understanding of traffic control devices. AAA Foundation for Traffic Safety, Falls Church, VA.

Hulbert, S., Beers, J., and Fowler, P. (1980). Motorists' understanding of traffic control devices II. AAA Foundation for Traffic Safety, Falls Church, VA.

International Standards Organization (1984). International standards for safety colors and safety signs. ISO 3864, Geneva, Switzerland.

Jabbar, A.S. and Naqvi, S.A. (1992). A study of road signs, *Derasat-Nafyseyah*, 2(3), 79–87.

Johansson, G. and Backlund, F. (1970). Drivers and road signs. *Ergonomics*, 13, 749–759.

Johansson, G. and Rumar, K. (1966). Drivers and road signs: a preliminary investigation of the capacity of car drivers to get information from road signs. *Ergonomics*, 9, 57–62.

Knoblauch, R.L. and Pietrucha, M.T. (1987). Motorists' comprehension of regulatory, warning and symbol signs. Vol. 1–3, Reports FHWA-RD-86-111–113, Department of Transportation, Washington D.C.

Lauer, A.R. and McMonagle, J.C. (1955). Do road signs affect accidents? *Traffic Q.*, 3, 322–329.

Ministry of Transport (1927). Manual and specification for the manufacture, display and erection of road markings and signs, Washington, U.S.

Ogden, M.A., Womak, K.N., and Mounce, J.M. (1990). Motorist comprehension of signing in urban arterial work zones. *Transp. Res. Rec.*, 1281, 127–135.

Pline, J.L. (1992). *Traffic Engineering Handbook*, 4th ed. Institute of Transportation Engineers. Prentice Hall, Upper Saddle River, NJ.

Richards, S.H. and Heathington, K.W. (1988). Motorist understanding of railroad highway grade crossing traffic control devices and associated traffic laws. *Transp. Res. Rec.*, 1160, 52–59.

Riegelneier (1942). Rehabilitating signs. *Proc. ITE*, 1942.

Road and Transportation Association of Canada (1966). Manual on uniform traffic control devices for Canada. Ottawa, Canada.

Sanders, J.H., Kolsrud, G.S., Jr., and Berger, W.G. (1973). Human factors countermeasures to improve highway–railway intersection safety. Report DOT-HS-800-888, U.S. Department of Transportation.

Shinar, D., Dewar, R., Summala, H., and Zakowska, L. (1999). Highway traffic sign comprehension: a cross-cultural study. *10th Int. Conf. Traffic Saf. Two Continents*, Malmo, Sweden, 67–69.

Standard Association of Australia (1960). Australian standard rules for the design, location, erection and use of road traffic signs and signals. Australia: Sydney.

Summala, H. and Hietamäki, J. (1984). Drivers' immediate responses to traffic signs. *Ergonomics*, 27(2), 205–216.

Summala, H. and Näätänen, R. (1974). Perception of highway traffic signs and motivation. *J. Saf. Res.*, 6, 150–154.

United Nations Economic Commission for Europe (1968). Convention on road signs and signal. Inland Transport Committee, U.N., Vienna.

United Nations Economic Commission for Europe (1974). European rules concerning road traffic signs and signals. European Conference of Ministers of Transport, Geneva. http://www.unece.org/trans/conventn.

United Nations Economic Commission for Europe (1995). Convention on road signs and signal done at Vienna on November 1968 Amendment 1*, Inland Transport Committee, U.N., Vienna.

Van Gelder, W., Rankin, W.W., and Baerwald, J.H. (1992). Traffic signs and markings. In Pline, J. (Ed.), *Traffic Engineering Handbook*, 5th ed. Washington, D.C.: Institute of Transportation Engineers, 411–452.

Wolff, J.S. and Wogalter, M.S. (1998). Comprehension of pictorial symbols: effect of context and test method. *Hum. Factors*, 40, 173–186.

Womack, K.N., Guesman, P.K., and Williams, R.D. (1981). Measuring effectiveness of traffic control devices: an assessment of driver understanding. Texas Transportation Institute, College Station, Texas.

Wright, P. (1996). *Highway Engineering*, 6th ed. Wiley, New York, 330–336.

11 Comprehension of Signs: Driver Demographic and Traffic Safety Characteristics

Hashim Al-Madani

CONTENTS

11.1 Sign Comprehension in Relation to Driver Origin ... 170
  11.1.1 Origin and Education ... 171
  11.1.2 Origin and Age ... 172
  11.1.3 Origin and Experience ... 172
11.2 Sign Comprehension in Relation to Driver Gender and Marital Status ... 172
  11.2.1 Gender in Relation to Age ... 173
  11.2.2 Gender in Relation to Experience ... 173
  11.2.3 Marital Status ... 173
  11.2.4 Marital Status in Relation to Experience ... 174
11.3 Sign Comprehension in Relation to Driver Age and Experience ... 174
  11.3.1 Driver Age and Experience ... 175
  11.3.2 Driver Experience ... 175
11.4 Sign Comprehension in Relation to Driver Education and Income ... 176
11.5 Sign Comprehension in Relation to Traffic and Safety Characteristics ... 177
  11.5.1 Accident Involvement ... 177
  11.5.2 Driver Lifetime Accident Involvement ... 177
  11.5.3 Driver At-Fault Accident Involvement ... 178
  11.5.4 Experience-Per-Accident Ratio ... 178
  11.5.5 Seat Belt Usage ... 179
  11.5.6 Citations ... 179
11.6 Combined Effect of Driver Socioeconomic and Safety Characteristics on Sign Comprehension ... 179
11.7 Conclusions and Implications ... 181
References ... 182

0-415-31086-5/04/$0.00+$1.50

This chapter covers drivers' comprehension of signs according to driver origin, age, gender, marital status, educational background, experience, accident involvement, and citation involvement. Furthermore, the direction, magnitude, and significance of each of these parameters when they are jointly considered are also presented in this chapter.

11.1 SIGN COMPREHENSION IN RELATION TO DRIVER ORIGIN

In Texas, Hawkins et al. (1993) observed particular problems in understanding traffic signs by non-Anglo–Saxon drivers such as Hispanics, African Americans, and those who did not speak English as their first language. Non-English-speaking drivers misunderstood around one third of signs. Al-Madani and Al-Janahi (2002a) found a driver's origin to be significantly related to his comprehension of signs. Western drivers are more knowledgeable in comprehending posted signs when compared with drivers from the Arab countries, Asia, Africa, and Oceanic countries because these countries established means of education regarding traffic signs a long time ago compared with many other countries. African and Oceanic drivers are the next most knowledgeable. Drivers from Arab states face difficulties in comprehending signs properly when compared with Western, Asian, African, and Oceanic drivers (Al-Madani and Al-Janahi, 2002a, b).

Nevertheless, even American drivers face difficulties with some signs. Ogden et al. (1990) found several work zone signs to be poorly comprehended by drivers in America. For example, signs indicating "Advance Construction" and "No Center Turn Lane" were comprehended by at most 42% of drivers. Drivers in the Middle East comprehend some signs even better than those in the U.S. The sign indicating "Slippery When Wet" was identified by more than 84% of drivers in the Middle East (Al-Madani and Al-Janahi, 2002a, b) but was comprehended by only 41% of drivers in America and Canada (Dewar et al., 1994).

Similarly, the "Lane Reduction" sign was comprehended by 86% of drivers in the Middle East (Al-Madani and Al-Janahi, 2002a) compared with 78% in Texas (Ogden et al., 1990). Galer (1980) found that 21% of truck drivers did not understand "Head Room Hazard" signs; 26% of drivers in the Arab world comprehended the sign correctly (Al-Madani and Al-Janahi, 2001). The "Lane Ends" symbol sign was correctly identified by 78% of Houston, Texas, drivers (Ogden et al., 1990) and 86% of drivers in the Arab region (Al-Madani and Al-Janahi, 2001). Al-Yousifi (1999, 2002) found that 85% of drivers in France, Kuwait, Bahrain, and Iran correctly identified the sign indicating "Road Narrows on Left Ahead," but only 80% of drivers in the Arab world correctly identified it (Al-Madani and Al-Janahi, 2001).

Arab, Asian, and Western drivers identified one third of regulatory and warning signs with equal success. Table 11.1 lists these signs. The regulatory signs, which were equally understood by drivers, were mostly those indicating prohibitory actions such as entry, turning, stopping, waiting, or passage (Al-Madani and Al-Janahi, 2001). Signs carrying pictorial symbols (i.e., symbol picture of a vehicle, bridge, pedestrian, etc.) were underrepresented (Al-Madani and Al-Janahi, 2001).

TABLE 11.1
Differences in Sign Comprehension among Arab, Asian, and Western Drivers

Sign Type	Signs Understood Equally Well by Arabs, Asians, and Westerners	Signs Poorly Understood by Arabs and Asians Compared with Westerners	Signs Poorly Understood by Arabs Compared with Westerners
Regulatory sign	No pedestrian passage Turn left Maximum speed limit No stopping No right turn No waiting No entry for vehicular traffic	End of prohibition of overtaking Priority to oncoming traffic Closed to automobiles End of prohibition of goods vehicles Keep right	Axle weight limit Mini roundabout Maximum height limit No U-turn Turn left ahead Vehicles may pass either side
Warning sign	Dual carriage with 3 lanes — right-hand lane is closed Road narrows on left ahead	Staggered junction Diversion to opposite Other danger plate indicates nature of danger	Traffic merges from left Road narrows from both sides Slippery road Dual carriage way ends Hump bridge ahead

Source: Adapted from Al-Madani, H.M.N. and Al-Janahi, A.R., Proc., Traffic Safety on Three Continents, Moscow, Swedish National Road and Transport Research Institute, Sweden, 2001.

Although many signs are comprehended well by Western drivers, the same cannot be said of Arab and Asian drivers. Western drivers identified almost one third more regulatory and warning signs than did Arab and Asian drivers. Warning and regulatory signs were represented significantly within the signs poorly comprehended by Arab and Asian drivers compared to Western drivers. When the signs that were comprehended equally by Arab, Asian, American, and European drivers are considered, symbolic pictorial signs are overrepresented here (Table 11.1, first column). An additional one third of the signs were understood significantly less well by Arab drivers compared with Western drivers (Table 11.1, last column).

In general, many of the signs that were comprehended well by drivers are included among the signs better understood by Westerners compared to Arabs. Arab drivers found warning signs more difficult to comprehend compared to Western drivers (Al-Madani and Al-Janahi, 2001). These findings raise a serious concern about the adequacy of such signs to Arab and Asian drivers in the Middle East.

11.1.1 Origin and Education

Al-Madani and Al-Janahi (2002a) compared drivers' comprehension of signs for those with high educational backgrounds (i.e., those with at least an undergraduate degree) and different nationalities in order to compare the drivers on a more equitable basis. This is because the Western drivers in the studied countries have higher educational backgrounds than others. The findings confirmed that Western drivers were better in comprehending signs than other drivers. Drivers' correct identifications of the warning signs were 54.5, 57.8, 69.4, 80.1, and 69.1% for Arabian Gulf, other Arabs, Asian, Western, Australian, and African drivers, respectively. Similarly, the scores in the regulatory signs were 54.1, 58.8, 62.8, 69.4, and 62.3%. All the nationality groups (except Australians and Africans) that differed significantly from each other were at a level of significance less than 0.05 (Al-Madani and Al-Janahi, 2002b).

11.1.2 Origin and Age

Europeans and Americans were significantly better than Asian and Arab drivers in comprehending signs in all the age categories except for the youngest (i.e., those under 24 years) (Al-Madani and Al-Janahi, 2002b).

11.1.3 Origin and Experience

With Asian drivers, experience was not associated significantly with comprehension of signs. However, drivers from other nations with more than 10 years of experience comprehended signs significantly better than less experienced drivers (Table 11.2). American and European drivers' comprehension of signs was significantly better than that of drivers from Arab, Asian, African, and Oceanic countries with at least 10 years of experience. Less experienced American and European drivers had significantly better comprehension than Arab drivers. The results possibly reflect the level of traffic education in these countries (Al-Madani, 2000; Al-Madani and Al-Janahi, 2002a).

11.2 SIGN COMPREHENSION IN RELATION TO DRIVER GENDER AND MARITAL STATUS

Mannering (1993) strongly recommended studies comparing men and women in the field of driver information processing. Male drivers' understanding of traffic signs was significantly better than that of female drivers (Al-Madani and Al-Janahi,

TABLE 11.2
Drivers' Percentage Comprehension of Signs Based on Origin and Driving Experience

	Years of Driving Experience[a]			
Nationality Group	**<5**	**5–10**	**10–15**	**>10–15**
Arab Gulf	51.2 (653)	52.4 (724)	54.1 (657)	51.0 (296)
Arabs	53.5 (177)	54.8 (200)	59.0 (326)	58.5 (251)
Asians	62.8 (151)	62.8 (139)	64.1 (204)	65.0 (67)
ECU and U.S.	67.4 (17)	65.3 (22)	73.2 (62)	76.7 (109)
African and Oceanic	63.3 (11)	59.6 (10)	62.3 (11)	79.3 (21)

[a] Numbers in parentheses represent sample sizes.

Source: Adapted from Al-Madani, H.M.N. and Al-Janahi, A.R., *Transp. Res. F*, 5, 63–76, 2002.

2002b). However, one might argue that such a significant difference between male and female drivers might be limited to drivers in certain parts of the world and cannot be generalized. To check this, Al-Madani and Al-Janahi (2002a) compared European and American male and female drivers of high educational backgrounds (with at least undergraduate degrees). Although the male drivers' comprehension of signs was better than that of the females, statistical tests did not show any significant impact of driver gender on comprehension of signs.

The average understanding of posted signs for European and American male drivers was 73% and for female drivers was 67% (Al-Madani and Al-Janahi, 2002a). Nevertheless, in Texas, men were better than women in correctly identifying almost one fifth of warning signs (Hawkins et al., 1993). Male drivers' better knowledge of signs may be explained by higher and wider exposure rates to the signs than those of female drivers. Besides covering less mileage annually, females are less exposed to signs other than those usually employed in urban areas (Al-Madani and Al-Janahi, 2002b).

Contrary to this, Luoma and Rama (1998) found recall of speed limit signs not to be affected by drivers' sex. Richards and Heathington (1988) also observed no significant difference between male and female drivers in their understanding of signs in the U.S. Hawkins et al. (1993) found that women better understood regulatory signs or pavement markings. In summary, clear differences exist between the previously mentioned studies' findings regarding differences between males and females in their understanding capabilities related to traffic signs. Exposure rate difference between the two genders might be, among other factors, an important reason behind such differences.

11.2.1 Gender in Relation to Age

When a driver's age is considered, male drivers' comprehension of signs is significantly better than females up to the age of 44. No significant difference between the two exists after that (Al-Madani and Al-Janahi, 2002b). Schmidt (1982) found typical observers of speed limits to be females aged between 40 and 60 years and typical nonobservers to be males between 25 and 40 years.

11.2.2 Gender in Relation to Experience

Experienced male drivers comprehend signs significantly better than female drivers in various driving experience categories. Female drivers' comprehension of signs does not improve significantly with driving experience. On the other hand, male drivers with more than 10 years of experience were found to have significantly better comprehension than less experienced ones. Once again, these can be simply explained by higher and wider exposure rates of male drivers compared with those of female drivers (Al-Madani and Al-Janahi, 2002a).

11.2.3 Marital Status

Comparison of single and married drivers requires considerable care because age and exposure-related parameters need to be considered. Age and years of experience distribution of married drivers does not match with that of single drivers in many countries (Al-Madani and Al-Janahi, 2002b). Although the majority of drivers in the age group of 16 to 24 years in the Middle East are single, the majorities in the older age groups are married. Peck et al. (1971) found that male and female single drivers have higher crash rates than married drivers of the same age. On the other hand, Evans (1991) stated that part of the difference between the two may be due to difference in the amount of driving (i.e., exposure rate). Married drivers, regardless of their age and experience, generally comprehended posted signs significantly better than single ones (Al-Madani and Al-Janahi, 2002b). However, this is not true when the two are compared for similar age groups. In fact, the differences are highly insignificant in all the age groups. Age is, therefore, the prime contributor to drivers' comprehension of signs, not marital status. Married drivers' greater social and economic responsibilities compared to those of single drivers do not lead to better comprehension of signs (Al-Madani and Al-Janahi, 2002b).

11.2.4 Marital Status in Relation to Experience

Although married novice drivers are better than single novice ones in understanding signs, unmarried experienced drivers are as good as married experienced ones (Al-Madani and Al-Janahi, 2002a). It is quite important to mention that this does not mean that novice unmarried drivers are comparable with experienced unmarried ones or that novice married ones are comparable with experienced married drivers. Single and married drivers with fewer than 10 years of experience understand signs significantly less well than those with more than 10 years of experience (Al-Madani and Al-Janahi, 2002a). European and American single and married drivers of higher educational qualifications have statistically equal comprehension of signs. The average understandings of posted signs for European and American single and married drivers are 73% and 69%, respectively.

11.3 SIGN COMPREHENSION IN RELATION TO DRIVER AGE AND EXPERIENCE

Otani et al. (1992) emphasized that not all signs produce differences due to age. For example, Hawkins et al. (1993) observed no significant relation between age and misunderstanding of over two thirds of tested regulatory and warning signs and pavement markings. They also found that older drivers misunderstood 18% of signs; young drivers misunderstood 12%. Similarly, Dewar et al. (1994) found no differences with respect to age in 61% of signs. However, older drivers are poorer than younger ones in understanding for the remaining 39% of examined signs. Surprisingly, Al-Sharea (1988) found that younger drivers perform much better than older drivers in traffic safety knowledge in Saudi Arabia. This could be due to better education of younger drivers in that country.

Hawkins et al. (1993) found at least 4 out of 31 tested signs to be misunderstood by many elderly drivers in the U.S. and also observed that comprehension of signs increases significantly with age only for speed zone signs. Richards and Heathington (1988) concluded that very young (under 19 years) and old (over 54) drivers in the

U.S. have difficulty in understanding railroad grade crossing traffic signs and signals. Luoma and Rama (1998) found recall of speed signs not to be affected by drivers' age. Otani et al. (1992) found that drivers over 60 years old are more at risk of ignoring warning signs. Moreover, Hofner (1982) investigated factors associated with compliance and noncompliance with traffic regulations, such as observing regulatory signs and safe driving practice. His results contradicted, at least with regard to age, the findings of Otani et al. (1992). Hofner found typical noncompliers, who tended to take risks, were between 45 and 55 years old.

The results of Al-Madani and Al-Janahi (2002a) revealed that drivers' comprehension of posted regulatory and warning signs, as a group, generally increases with age. However, differences between various age groups showed that drivers in the younger age group (16 to 24 years) comprehended significantly less well than those in the older groups (35 to 44 and over 44 years). Allen et al. (1980) also observed declines in knowledge of symbolic signs with age. Hulbert and Fowler (1979) found significant differences in comprehension, in relation to age, in seven out of eight tested signs. Levels of comprehension were 70% for drivers aged less than 24 years; 79% for those 24 to 50 years old; and 72% for those over 50 years old. In a follow-up study, Hulbert et al. (1980) found older drivers (over 50 years) comprehended signs significantly less well than other age groups. The correct response rate of the older drivers was below 60%.

One may conclude that drivers of different age groups understand many signs equally; many others do not. Younger drivers understand the majority of signs better than older drivers.

11.3.1 Driver Age and Experience

In most of the motorized countries of the world, age and experience are highly positively correlated (Hawkins et al., 1993; Mannering, 1993), which is also true in many Arab countries (Al-Madani and Al-Janahi, 2002b). Either of the two would be sufficient for analysis of drivers' understanding of signs. Middle-aged (35 to 44 years) and older (45 to 55 years) drivers of different experience groups understand the signs significantly better than younger ones in the Arab world (Al-Madani and Al-Janahi, 2002a). Experience has no significant influence in improving drivers' comprehension of regulatory or warning signs when the drivers reach the age of 45 years and over (Al-Madani and Al-Janahi, 2002b). In other words, novice and experienced drivers (45 years of age or older) comprehend the signs equally. It appears that knowledge gained about signs before the age of 45 or later, along with the information comprehended about signs while provisionally driving, does not improve further with years of driving experience (Al-Madani and Al-Janahi, 2002b).

11.3.2 Driver Experience

Hawkins et al. (1993) observed that more than 10% of regulatory and warning signs are misunderstood by drivers with low levels of experience. Al-Madani (2000) compared drivers' understanding of posted signs in the Middle East for various experience groups of drivers, regardless of age. Highly experienced drivers (i.e., those with at least two decades of experience) understood warning and regulatory signs significantly better than novice drivers did (i.e., those with 5 years of experience, at most). Richards and Heathington (1988) observed significant differences between novice and experienced American drivers. They also observed that drivers covering fewer than 5000 miles a year tend to perform worse than other drivers in comprehension of signs. Furthermore, married and single drivers with more than 10 years of experience comprehended the signs better than less experienced ones (Al-Madani and Al-Janahi, 2002b).

11.4 SIGN COMPREHENSION IN RELATION TO DRIVER EDUCATION AND INCOME

Comprehension of signs increases with a driver's level of education. The differences among the various educational groups indicate that drivers with low educational backgrounds (i.e., diploma holders or lower) understand posted signs significantly less well than those with higher educational backgrounds (Al-Madani and Al-Janahi, 2002b). Furthermore, drivers with lower levels of education in general misunderstand 15% of signs (Hawkins et al., 1993) and also show poor comprehension of grade crossing control devices (Richards and Heathington, 1988). Comprehension of signs increases significantly with education (Table 11.3) for various experience categories of drivers (Al-Madani and Al-Janahi, 2002a). However, in some cases, the opposite is true because comprehension does not increase significantly with experience for those with high qualifications (i.e., master's and Ph.D. holders) but increases for those holding a diploma or a B.Sc. degree (Al-Madani and Al-Janahi, 2002a). Wealthy drivers understand signs significantly better than poor drivers (Al-Madani

TABLE 11.3
Drivers' Percentage Comprehension of Signs Based on Education and Driving Experience

	Years of Driving Experience[a]			
Educational Background	**<5**	**5–10**	**10–15**	**>15**
Below secondary	49.1 (132)	49.7 (113)	53.4 (110)	50.5 (82)
Secondary, but no diploma	52.8 (328)	54.0 (306)	55.9 (250)	55.5 (106)
Diploma holders	55.4 (226)	51.3 (267)	56.1 (222)	59.7 (151)
B.Sc. degree	55.2 (255)	57.5 (347)	59.2 (526)	61.7 (261)
Higher studies	60.0 (42)	62.5 (51)	64.0 (146)	63.2 (132)

[a] Numbers in parentheses represent sample sizes.

Source: Adapted from Al-Madani, H.M.N. and Al-Janahi, A.R., *Transp. Res. F*, 5, 63–76, 2002.

and Al-Janahi, 2002b). Moreover, high-income drivers comprehended signs significantly better than low-income drivers at every level of driving experience. This may be partly because of better chances of traffic education for wealthy drivers.

11.5 SIGN COMPREHENSION IN RELATION TO TRAFFIC AND SAFETY CHARACTERISTICS

11.5.1 Accident Involvement

The use of engineering improvement programs for signs has been shown to reduce accidents considerably (Al-Madani and Al-Janahi, 2002b). Many accidents occur because the driver is suddenly confronted by the unexpected. The driver should therefore be warned by means of traffic signs, as much as possible, for any abnormal driving situations ahead (Pignataro, 1973). Signs are essential where special regulations apply at specific places or at specific times and where hazards are not self-evident. According to the U.S. Department of Transportation's annual report on highway safety improvement programs (1989; 2001), improvement in traffic signs at intersections can lead to a 34% reduction in fatal accidents and a 93% reduction in injury-related accidents in the U.S.

In fact, the effectiveness in relation to benefit/cost ratio appears to be much higher compared to improvements in traffic channelization facilities such as painted or raised islands, sight distance, markings, illumination, and traffic signal improvements. Improvements in posted signs include their location and illumination. Schuster (2002) claimed a 30% reduction in accident costs on German motorways as a result of improvements in permanent traffic signs, considering speed, prohibition of overtaking, and various other message signs.

Although many studies have covered the influence of driver education on crash rates, none used an acceptable methodology to show that those who receive driver education have lower crash rates than those who do not (Evans, 1991). Brown et al. (1988), for example, found no reliable evidence of safety benefits or accident rate improvements (Patvin et al., 1988) from driver training. This may raise a question of concern: do drivers with lower accident frequencies or accident rates per years of driving experience have better knowledge of traffic signs?

Drivers who disregard plausible speed limits were more often involved in serious accidents and committed more traffic violations than those who observed the speed limits (Richards and Heathington, 1988). Furthermore, Otani et al. (1992) found higher risk perception to be related to lower likelihood of ignoring signs. Hofner (1982) found that typical noncompliers tend to take risks. In America, drivers performing more safely at crossings were also observed to be interpreting traffic control devices more correctly (Sanders et al., 1973).

11.5.2 Driver Lifetime Accident Involvement

Al-Madani et al. (1996) studied the influence of drivers' comprehension of regulatory and warning signs on their involvement in accidents. Their results revealed that no significant differences between any two groups of drivers' lifetime accident involvement (at fault or otherwise) were observed among the group of signs including warning messages. A significant difference was observed between drivers with no accidents and those with high numbers of accident involvement in terms of sign comprehension. Furthermore, they tested the influence of drivers' accident involvement on seven individual signs that they thought were related to the safety of drivers.

Drivers of different accident involvement levels comprehended five of the signs equally. These signs were: priority to oncoming traffic, diversion to opposite, slippery roadway, mini roundabout, and pedestrian passage. Drivers with no accident involvement comprehended the signs indicating staggered junction and roadway narrowing better than those with high accident frequency. Drivers involved in at least three accidents comprehended warning and regulatory signs much less than drivers with fewer accident involvement records. Apart from the few exceptions mentioned earlier, neither accidents nor lifetime accident involvement proved to be associated with drivers' sign comprehension (Al-Madani, 2000, 2001).

11.5.3 Driver At-Fault Accident Involvement

To understand drivers' accident involvement better in relation to their comprehension of signs, one may consider the number of accidents in which the drivers were at fault. Regulatory and warning signs are equally understood by drivers with various numbers of at-fault accident involvement (Al-Madani, 2000). This means that drivers' better understanding of posted regulatory and warning signs has no significant influence on reducing accidents that are their faults. In other words, drivers with high accident involvement are statistically as good in recognition of traffic signs as those with no accident involvement. A similar result was observed even when age was considered. In other words, no significant difference exists in comprehension of signs between any two accident categories (i.e., 0, 1, 2, or 3 and more accidents) in any age group. Risk-taking drivers are as good as cautious ones in understanding traffic signs (Al-Madani, 2000).

11.5.4 Experience-Per-Accident Ratio

In order to consider the combined effect of drivers' involvement in accidents and their driving experience on their understanding of posted signs, experience-per-accident ratio was used to take care of the cumulative effect of the accidents with years of experience on drivers' comprehension of signs (Al-Madani, 2000). It is not fair, for example, to compare the understanding level of a driver who was involved in a single accident in a span of 20 years of driving to that of one with 5 years of experience. The results, considering drivers' total accident involvement and accidents due to their own faults, indicated no significant differences between any two groups. However, drivers in the groups with high accident rates per experience ratios were generally better than those in the low accident rate groups (Al-Madani, 2000).

In conclusion, one may say that traffic signs promote traffic safety by providing orderly movement of all road users and warning users against any potential hazards; however, drivers' comprehension of signs is not associated with their accident involvement. This does not mean that authorities should give less priority to the improvement of roadway signs or to driver education programs because, as seen throughout this book, signs fulfill other needs, as well.

11.5.5 Seat Belt Usage

Drivers who always use their seat belts comprehend regulatory and warning signs significantly better than those who sometimes or occasionally use them. Similarly, those who sometimes use them comprehend signs better than those who occasionally use them. As drivers' awareness of posted signs increases, so does their seat belt usage (Al-Madani, 2000).

11.5.6 Citations

Drivers' understanding of the speed limit sign is usually high; for example, in Bahrain only 9% of drivers do not know the meaning of this sign. In other Arabian Gulf countries, approximately 23% do not know the meaning well (Al-Madani, 2000). Typical observers of speed limits are females aged between 40 and 60 years and typical nonobservers are males between 25 and 40 years (Schmidt, 1982). Drivers with various speed citations understand regulatory and warning signs equally. In other words, drivers with good understanding of signs are not necessarily involved in fewer speed citations compared to those with less understanding (Al-Madani, 2000). In contrast to this, Hofner (1982) found that those who complied with traffic signs received fewer traffic fines.

In contrast to the speed limit sign, the majority of the drivers from the Arab world poorly comprehend signs indicating "No Stopping" (Al-Madani, 2000). In the U.S. and Canada, 98% of the drivers correctly identified the "No Parking" sign (Dewar et al., 1994). Drivers with a high number of parking citations comprehend regulatory and warning signs the least and those with no citations comprehend them the best; however, compared with drivers with a different number of parking citations, statistically no two groups differ significantly (Al-Madani et al., 1996).

11.6 COMBINED EFFECT OF DRIVER SOCIOECONOMIC AND SAFETY CHARACTERISTICS ON SIGN COMPREHENSION

Several studies by Al-Madani (2001) and Al-Madani and Al-Janahi (2002a) analyzed the joint influence of drivers' socioeconomic and safety characteristics on their comprehension of regulatory and warning signs through regression analysis. Such studies are useful in predicting the direction and magnitude of the association of each predictor on drivers' comprehension of signs when they are jointly considered. They are particularly important when colinearity exists between some of the considered variables, as in the case of age and experience or of age, income, and education. Only the predictor variable with the highest association is included. The others become unbalanced and are removed from the regression.

Richards and Heathington (1988) observed that the three factors jointly affecting drivers' comprehension are education, age, and driving experience. Al-Madani and

Al-Janahi (2001) carried out modeling for drivers' recognition of signs. Arabs in general and drivers from Arabian Gulf countries in particular are negatively associated with comprehension of signs. On the other hand, Europeans and Americans are positively associated with it. The regressed data also revealed that drivers' comprehension increases with age, monthly income, and years of education — squared. Male drivers showed a positive association with comprehension of signs; married drivers proved to be insignificant through all the analyses (Al-Madani, 2001).

In addition to the earlier predictors, Table 11.4 shows further regression results for drivers' comprehension of 28 regulatory and warning signs, considering drivers' accident-per-experience rate (Al-Madani and Al-Janahi, 2002a). Drivers' monthly income and years of education are significantly related to their understanding of posted signs. On the other hand, age is not associated with comprehension in the presence of other predictors. The quadratic terms of these variables showed none, except income, to be significant. Because age and years of education are highly correlated, none of the modeled variables showed that both taken together as linear predictors were significant (Al-Madani, 2001). Regressing such data is usually accompanied by low R-square values and high F-test results, probably because of

TABLE 11.4
Coefficient Estimates and Standard Errors of Predictors for Drivers' Comprehension of Signs

Parameter	Coefficients	Significance
Constant	34.887	0.000
Nationality (GCC = 1; Arab = 2; Asian = 3; Westerner = 4; other = 5)	5.167	0.000
Age (linear)	UV[c]	UV
Age (quadratic)	UV	UV
Education in years (linear)	0.457	0.000
Education in years (quadratic)	UV	UV
Income[a] (linear)	3.850×10^{-3}	0.000
Income[a] (quadratic)	-4.847×10^{-7}	0.007
Sex (male)	4.495	0.000
Married (yes)	UV	UV
Accident rate[b]	UV	UV
Experience (years)	UV	UV
Model statistics	Adj. $R^2 = 0.248$; $F = 172.4$	

[a] In Bahraini dinars (B.D.); 1 B.D. = U.S. $2.67.

[b] (Accident + 1) per experience rate.

[c] UV = unbalanced variables.

Source: From Al-Madani, H.M.N., *Perceptual Motor Skills*, 92, 72–82, 2001.

the dispersion nature of the data, which is common in such studies (Al-Madani, 2001). Johansson and Backlund (1970), for example, observed such variation when they tested only five signs. Shinar et al. (1999) also found that sign recognition varied widely among the drivers (Al-Madani and Al-Janahi, 2002a).

In the presence of other variables, years of driving experience and drivers' total accident involvement per years of experience have no significant effect on drivers' comprehension of signs (Table 11.4). Similarly, drivers' at-fault accident involvement has no effect on sign comprehension, probably because of their correlation with other included variables. However, when comprehension of signs is regressed onto accident ratio alone, it proves to be highly significant. Moreover, when the variables are removed systematically, one at a time, the dominating variables are age, monthly income, and gender. The presence of any of these variables causes the accident rate-per-experience ratio of either type to become unbalanced. Likewise, years of experience proved to be significant when regressed alone and stays significant in the presence of other variables if origin or monthly income is not present (Al-Madani et al., 2002a). The predicted model is as follows (Table 11.4):

$$\text{Driver's comprehension of signs} = 34.89 + 5.17(O) + 0.46\,(E) + 3.85 \times 10^{-3}\,(I) - 4.9 \times 10^{-7}\,(I - Iav)^2 + 4.50\,(M)$$

where O is drivers' origin as categorized in Table 11.4; E is years of education; I is monthly income (1 B.D. = \$2.67 U.S.); Iav is average monthly income of the country; and M is gender type (positive for male and negative for female). Such models are usually not precise because of limitations in the technique used and because of dynamic changes in drivers' comprehension of signs with time. Other variables might affect this relation, as well, some of which have not yet been properly considered in driver behavior research. However, the equation may be used as a predictive or assessment tool as well as gain information on drivers' understanding process when comprehension of traffic signs is considered (Al-Madani et al., 2002a).

11.7 CONCLUSIONS AND IMPLICATIONS

Traffic signs enhance traffic safety by providing orderly movement of road users and warning them against potential hazards. However, neither a driver's accident involvement nor his accidents related to his experience are associated with comprehension of signs. In other words, good comprehension of posted signs does not reduce the driver's accident involvement or vice versa. However, this does not mean that agencies responsible for driver licensing and for developing driver manuals should not be adequately supported, because traffic signs fulfill other driving needs (e.g., navigational). Furthermore, drivers' good comprehension of signs does not mean less speed or fewer parking citations when compared to those with less comprehension capabilities.

Novice and young drivers who are under 24 years old comprehend signs less well than experienced and older drivers. Similarly, female drivers in the various experience categories comprehend signs less well than males. Married and single drivers comprehend them equally. Well-educated drivers and those with high incomes comprehend signs better than those with low levels of education and income. Western drivers of various educational backgrounds understand signs better than drivers of many other nationalities. When sociodemographic characteristics are jointly regressed, the dominating variables are age, income, origin, and gender type. However, experience may replace origin or income.

Traffic education, particularly in less developed countries when compared with developed ones, should be given greater attention in order to increase drivers' comprehension of roadway signs. Furthermore, as will be touched upon later in this book, a new generation of improved signing that suits the needs of drivers in motorized and less motorized countries is certainly needed.

REFERENCES

Allen, R.W., Parseghain, Z., and van Valkenburg, P.C.A. (1980). Simulator evaluation of age effects on symbol sign recognition. *Proc. Hum. Factor Soc., 24th Annu. Meet.*, Los Angeles: Human Factors Society.

Al-Madani, H.M.N. (2000). Influence of drivers' comprehension of posted signs on their safety related characteristics. *Accident Anal. Prev.*, 32, 575–581.

Al-Madani, H.M.N. (2001). Prediction of drivers' recognition of posted signs in five Arab countries. *Perceptual Motor Skills*, 92, 72–82.

Al-Madani, H.M.N. and Al-Janahi, A.R. (2001). Differences in traffic signs' recognition between drivers of different nations. *Proceedings of Traffic Safety on Three Continents*, Moscow, Swedish National Road and Transport Research Institute, Sweden.

Al-Madani, H.M.N. and Al-Janahi, A.R. (2002a). Assessment of drivers' comprehension of traffic signs based on their traffic, personal and social characteristics. *Transp. Res. F*, 5, 63–76.

Al-Madani, H.M.N. and Al-Janahi, A.R. (2002b). Role of drivers' personal characteristics in understanding traffic sign symbols. *Accident Anal. Prev.*, 34, 185–196.

Al-Madani, H.M.N., Al-Janahi, A.R., and Abdul Ghani, A.A. (1996). Motorists' conception of posted regulatory and warning signs. *Proc., 29th Int. Symp. Automotive Technol. Automation*, Florence, Italy, 131–138.

Al-Sharea, A. (1988). Behavior of the drivers and safety measures (in Arabic). *Proc., 3rd IRF Middle East Conference*, Saudi Arabia, 2, 205–216.

Al-Yousifi, A.E. (1999). Investigation of traffic signs to improve road safety. *Proc., 10th Int. Conf. Traffic Saf. Two Continents*. VTI Konferens 13H. Malmo, Sweden, 73–87.

Al-Yousifi, A.E. (2002). A closer look at our traffic signs. *Safety on Roads, 2nd Int. Conf.*, SORIC '02, Center for Transport and Road Studies, University of Bahrain, Paper no. E147, Bahrain.

Brown, I.D., Groeger, J.A., and Biehl, B. (1988). Is driver training contributing enough towards road safety? In Rothengatter, J.A. and De Bruin, R.A. (Eds.), *Road Users and Traffic Safety*. Assen/Masstricht, Netherlands: VanGorcum.

Dewar, R.E., Kline, D.W., and Swanson, H.A. (1994). Age differences in comprehension of traffic sign symbols. *Transp. Res. Rec.*, 1456, 1–10.

Evans, L. (1991). *Traffic Safety and the Driver.* New York: Van Nostrand Reinhold.

Galer, M. (1980). An ergonomics approach to the problem of high vehicles striking low bridges. *Appl. Ergonomics*, 11(1), 43–46.

Hawkins, H.G., Womak, K.N., and Mounce, J.M. (1993). Driver comprehension of regulatory signs, warning signs and pavement markings. *Transp. Res. Rec.*, 1403, 67–82.

Hofner, K-J. (1982). Causes of traffic violations, *Arbeiten-aus-dam-Verkehrspsychologischen-Institut*, 19(6), 47–58.

Hulbert, S. and Fowler, P. (1979). Motorists' understanding of traffic control devices. AAA Foundation for Traffic Safety, Falls Church, VA.

Hulbert, S., Beers, J., and Fowler, P. (1980). Motorists' understanding of traffic control devices II. AAA Foundation for Traffic Safety, Falls Church, VA.

Johansson, G. and Backlund, F. (1970). Drivers and road signs. *Ergonomics*, 13, 749–759.

Luoma, J. and Rama, P. (1998). Effects of variable speed limit signs on speed behaviour and recall of signs. *Traffic Eng. Control*, 39, 234–238.

Mannering, F.L. (1993). Male/female driver characteristics and accident risk: some new evidence. *Accident Anal. Prev*, 25, 77–84.

Ogden, M.A., Womak, K.N., and Mounce, J.M. (1990). Motorist comprehension of signing in urban arterial work zones. *Transp. Res. Rec.*, 1281, 127–135.

Otani, H., Leonard, S.D., Ashford, V.L., and Bushore, M. (1992). Age difference in perception of risk. *Perceptual Motor Skills*, 74(2), 587–594.

Patvin, L, Champagne, F., and Laberge–Nadeau, C. (1988). Mandatory driver training and road safety: the Quebec experience. *Am. J. Public Health*, 78, 1206–1209.

Peck, R.C., McBride, R.S., and Coppin, R.S. (1971). The distribution and prediction of driver accident frequencies. *Accident Anal. Prev.*, 2, 243–299.

Pignataro, L.J. (1973). *Traffic Engineering: Theory and Practice.* Upper Saddle River, NJ: Prentice Hall.

Richards, S.H. and Heathington, K.W. (1988). Motorist understanding of railroad highway grade crossing traffic control devices and associated traffic laws. *Transp. Res. Rec.*, 1160, 52–59.

Sanders, J. H., Kolsrud, G. S., Jr., and Berger, W.G. (1973). Human factors countermeasures to improve highway–railway intersection safety. Report DOT-HS-800-888, U.S. Department of Transportation.

Schmidt, L. (1982). Observance and transgression of local speed limits, *Arbeiten-aus-dam-Verkehrspsychologischen-Institut*, 19(6), 107–116.

Schuster, G. (2002). Variable or permanent traffic signs on motorways, *Safety on Roads: 2nd Int. Conf. SORIC '02*, Center for Transport and Road Studies, University of Bahrain, Paper no. E129, Bahrain.

Shinar, D., Dewar, R., Summala, H. and Zakowska, L. (1999). Highway traffic sign comprehension: a cross-cultural study. *10th Int. Conf. Traffic Saf. Two Continents*, Malmo, Sweden, 67–69.

U.S. Department of Transportation (1989). Annual report on highway safety improvement programs. FHWA, Washington D.C.

U.S. Department of Transportation (2001). Manual on uniform traffic control devices for streets and highways — MUTCD Millennium edition, Department of Transportation, FHWA, Washington D.C., http://mutcd.fhwa.dot.gov. and http://mutcd.fhwa.dot.gov/kno-overview.htm. Retrieved January 4, 2003.

12 Specific Design Parameters: VMS Part I

Luís Montoro, Antonio Lucas, and María T. Blanch

CONTENTS

12.1 Introduction 185
12.2 VMS and the Road User 186
12.3 General Considerations for VMS Design 187
12.3.1 Message Content 189
12.3.2 Text Messages on VMS 189
12.3.2.1 Message Length 190
12.3.2.2 Number of Rows 190
12.3.2.3 Message Format 190
12.3.2.4 Bilingual Messages 191
12.3.2.5 Abbreviations 192
12.3.2.6 Message Absence 192
12.3.3 Pictogram and Combined Messages on VMS 193
12.3.3.1 Two-Frame or Alternated VMS 194
12.3.4 Causes vs. Consequences: On VMS Reading Schemas 194
12.4 Conclusions and Practical Implications 196
Acknowledgments 197
References 197

12.1 INTRODUCTION

In the road environment, the main functions of Intelligent Transport Systems (ITS) are to promote road safety and improve traffic flow via effective communication to drivers. Variable Message Signs (VMS) are currently one of the most powerful and widely implemented tools for this matter. These new road signs are indeed flexible: a VMS can be on and off; serve different road functions (warn, regulate, advise); display different elements (pictograms, text, or both); and adopt different message formats. By displaying complex, real-time road information, VMS intend to contribute to a higher efficiency and integration on the road transport system.

Several drawbacks may appear, though. On the one hand, VMS normally operate on high-speed roads; displaying too much information may have negative effects (e.g., overload; distraction; anomalous reactions, e.g., drivers slowing down

0-415-31086-5/04/$0.00+$1.50

abruptly). On the other hand, even when the basic informative elements are on, some structural parameters must be adopted in order to preserve and ease regular interpretation of information on the part of drivers. The main goal of this chapter is to offer some considerations and design recommendations for better use of VMS on the road. Although this chapter focuses on road transport, many of the general issues addressed here will also be relevant to new technologies for rail, aviation, maritime, or pedestrian signing.

12.2 VMS AND THE ROAD USER

In 1971, Allen et al. presented a general classification of driver behavior pictured as a three-level hierarchy: (1) the highest level (navigation), comprising behaviors related to trip planning and route finding; (2) the intermediate level (guidance), focused on road and traffic situation management; and (3) the lower level (control), including steering and speed control (see Sagberg, 1999). Similarly, other traffic researchers have adopted a hierarchical structure of driving behavior in terms of strategic/planning, tactical/maneuvering, and operative/control tasks (Ranney, 1994). VMS can be implemented in order to influence performance at different task levels, each level with distinct information needs in terms of time and priority. At the planning level, information or directional VMS may operate to help road users select the best route. At the maneuvering level, VMS may indicate some restriction (overtaking restrictions, lane control, variable speed control) or what to do in the proximity of danger (speed advice, weather information, traffic intensity). At the operational level, VMS try to guide highly automatized control behavior and information processing (speed and direction control) in order to gain unexpected but needed conscious attention (e.g., on speed; see Chapter 13).

Like other types of signs described in this book, VMS must be designed and operated so that drivers perceive and comprehend the signs and behave in the way expected (i.e., are influenced). For a sign to influence, it must first be perceived; thus, aspects such as size and luminosity, visibility (day, night), color inversion, and panel height and position received the earliest considerations (Dudek and Huchingson, 1986; Fabre et al., 1986). In addition to perception and comprehension factors, influence depends mainly on the whole VMS system operation and reliability in the face of drivers: useful diversion strategies, accurate journey times, good estimations on distance to lines, common sense variable speed limits, and so on. In short, it depends on the VMS system's worth in terms of drivers' travel goals and needs during a long time (Arbaiza, 2000).

In this chapter, VMS design refers mainly to comprehension issues. Factors affecting comprehension include:

- Available time to process information.
- Information-processing facilities (e.g., number of elements forming the message).
- Signs' (pictograms') "affordance".
- Eliciting of common reading schemas.
- Role played by inference.

- Use of clear-cut road sign categories.
- Drivers' previous learning and/or experience (e.g., commuters).

The question of whether a sign also elicits the intended action by the driver is a difficult but crucial one. Lewis and Cook (1969) provide a nice example about the warning sign for falling rocks — normally very well understood by drivers. However, when asked about what they would do upon seeing this sign, half of drivers responded that they would speed up in order to avoid falling rocks and leave the endangered area as quickly as possible; the other half indicated that they would slow down in order to be able to avoid fallen rocks (see Riemersma et al., 1986). This example points to a defect in some traffic signs, whether fixed or variable; namely, they arouse the driver to do something but do not state explicitly what behavior is expected.

VMS are usually displayed by using two types of elements: pictograms and alphanumeric characters ("text"). Words are probably the most powerful way to communicate because verbal signs reach the largest and most subtle regions in the general semantic space (Eco, 1976/translated 1995). However, as mentioned earlier in this book, road signs are showed in a (physically) limited space and are normally seen in a few seconds; too many words may interfere with, overload, or distract from the driving task.

In addition, road sign systems tend to form a universal code beyond local or national culture and language; pictograms serve this goal better than words.

> Well-designed pictorials can quickly communicate concepts and instructions at a glance. Pictorials may also be useful to persons who cannot read printed verbal messages because of vision problems, inadequate reading skills or unfamiliarity with the language used in the message.
>
> **Wogalter et al., 1997, p. 531**

These text–pictogram divergences and preferences appear in VMS systems worldwide.

12.3 GENERAL CONSIDERATIONS FOR VMS DESIGN

VMS systems can improve road network mobility and safety via information to road users. This general goal implies two important subgoals concerning, on the one hand, the VMS within the road network and, on the other, the VMS in front of the road user. The first subgoal concerns VMS systemic operational issues. Two broad types of VMS actions might be distinguished: (1) *tactical actions* affecting only the stretch of the road where the VMS is placed (e.g., lane closures, and speed limits) and (2) *strategic actions* in which a message issued on a VMS affects roads other than the one where it is located (e.g., rerouting, and congestion warnings) (Nenzi, 1997). Road circumstances determine VMS purpose; its main goal is user comprehension and compliance (Table 12.1). Following distance to event, network, and flow characteristics, among others, a host of different parameters are considered such as type,

TABLE 12.1
Outline of (Possible) VMS Actions

Message Type	Function	Content
Regulatory	Lane allocation	Lane change/merge (white/yellow diagonal arrows; flashing or with separate flashers); lane closure (red crosses); available/free lane (green arrow pointing downward)
	Carriageway guidance	Closure (road, bridge/tunnel, exit); mandatory exit (diversion)
	Speed control	Speed limit; speed funneling; speed harmonization
	Regulations	Restrictions of use (dedicated lanes for target groups: buses, trucks, carpools, etc.); temporary prohibitions; passing prohibitions (all vehicles); overtaking prohibitions (heavy vehicles); use of snow tires/chains
	End of (temporary) restrictions/limitations	
Danger warning	Immediate warning for weather conditions (close ahead)	Fog, rain, snowfall, reduced visibility, high winds, snow, black ice, slippery road, spillage
	Immediate warning for traffic status (close ahead)	General warning (plus additional subscription); congestion/line; incident/accident; roadwork; road closure ahead; oncoming vehicle; bridge closed for road traffic
	Warning for road status	Narrowing road ahead
Informative	Advance warning	Traffic status (further ahead or on another motorway section); weather conditions (further ahead); road status (further ahead)
	Implicit advice	Suggested route/itinerary (rerouting); suggested/optional exit; network performance after a decision point (travel times or extent of congestion on more than one route after on-coming junction); recommended maximum speed
	Driver comfort	Temporary available/free lane ahead (tidal flow lane, emergency stopping lane, etc.); recommended maximum speed
	Miscellaneous	Should be limited: only when relevant for driving conditions and not compromising traffic safety

Source: From WERD/DERD, Final version 3.0, Spring 2000. West European Road Directors (WERD), Deputy European Road Directors (DERD), p. 21, 2000.

number and format of VMS units, message prioritization, and operation type (manual, automatic) (SETRA, 1994; Friedrich et al., 1999).

The second subgoal concerns VMS design issues. A traditional classification will be followed according to information elements that may be exhibited (text, pictogram, or both). VMS display parameters differ in terms of number of pictograms (none, one, two); number of rows (between two and five; normally, three); characters per row (between 10 and 20 characters or more; normally about 16 to18); information distribution (pictogram on the left, text beside; pictogram above, text below; etc.); and supplementary features (e.g., flashing lights). Such variations may enrich message exhibition but also suppose a wide path for road sign heterogeneity. The problem may extend further if, when dealing with long messages, one considers the possibility of partitioning the message into two panels (see Section 12.3.2.2).

12.3.1 Message Content

One may think of VMS as telegraphs on the road. A message must be transmitted, but using too much information will be expensive in terms of cognitive resources, so it must be optimized toward a minimum. For example, no more than three or four categories for describing traffic states (e.g., light, slow, or heavy traffic and congestion) or delay levels (e.g., 15, 30, 45 minutes; 1 hour or more) should be displayed (Beccaria et al., 1991). Labels for describing events and situations should be carefully selected according to driver types and road situations (e.g., commuters prefer locations rather than exit numbers and "accident" is preferred to "crash") (Dudek, 2002; Beccaria et al., 1991). Message content (purpose), length, and load are somehow related (e.g., danger warning messages should be short and consequences immediate).

Three basic elements may fit the general informational requirements: (1) telling drivers *what is wrong ahead* (causes); (2) *where* (location, distance); and (3) *what should they do about it* (consequences). Dudek (2002, pp. 5–7) has carefully listed the basic VMS elements required to make a fully informed driving decision:

- Incident/roadwork descriptor (informing of the unusual situation).
- Incident/roadwork location (its location).
- Lanes affected.
- Closure descriptor.
- Location of closure.
- Effect on travel (e.g., diversion recommended).
- Audience for action.
- Action (what to do).
- Good reason for following the action.

Reasons and problems may be many and it is important to apply generic (not specific) labels to describe them. Consequences (situations, actions) may be reduced to a few (e.g., lane or road closure, variations in speed, diversion, and increasing alert) and should be absolutely transparent (i.e., concrete and specific).

12.3.2 Text Messages on VMS

Recommendations for text-based VMS design are important in the U.S. because text is the predominating symbol there (in the sense given by Krampen, 1983; see Chapter 13). Following the Vienna Convention, informative messages within the European area may use pictograms and text.

12.3.2.1 Message Length

Limiting message length also depends on VMS capacity in terms of rows and characters per row. The three possibilities for long text messages are: (1) using standard abbreviations, (2) eliminating redundant words, and (3) displaying the message in two panels (alternate). Travel speed and activity on the road (e.g., congested flow and sign confluence), which affect reading time, influence message length. "The legibility distance must allow drivers adequate time to read the sign 'twice' whilst attending to the driving task. Drivers should be able to finish reading before their eyes are diverted more than 10° from the road ahead" (Beccaria et al., 1991; pp. 4–15).

SETRA guidelines (1994) suggest a simple formula for evaluating the reading time: $t = N/3 + 2$ (seconds), where N is the number of reading units (words, numbers, symbols). Clearly, reading times improve with commuters and familiar messages. Following this formula, and supposing the VMS is located 7.5 m above the ground, a maximum reading angle of 10° (the driver's eyes are at 1.20 m above the ground), a maximum speed of 130 km/h, characters' height of 400 mm, and normal visual acuity, the initial reading distance is 200 m and the driver has 4.5 sec available (164 m) for reading the message completely. This means that, following the given formula, no more than seven units (words) should be considered (SETRA, 1994; Appendix 1).

Dudek (2002) provides an extensive set of recommendations concerning the number of units and speed in several road situations (day/night, vertical/horizontal curves, rain/fog, etc.; see module 7), advising no more than seven words at 65 mph and eight words at 55 mph. VMS are normally installed in high-speed roads (up to 120 to 130 km/h) and most European road administrations advise between four and seven information units (Beccaria et al., 1991; MacLaverty et al., 1998a). Moreover, between 40 and 80% of drivers will "moderately" speed up to 20 to 40 km/h over the limit (Cooper, 1997; MASTER, 1999). Too long or too complicated messages may lead some drivers to slow down (dangerously) in order to read them.

12.3.2.2 Number of Rows

Studies comparing comprehension levels of a fixed number of words distributed along a varying number of lines found that, compared with a four-line, two-words-per-line allocation, a two-line message with four words per line offered maximum level of comprehension. A maximum of six words and two lines of text appear to offer the optimum level (Hitchins et al., 2001). Similarly, European recommendations indicate three lines as the maximum for messages concerning road information and preferably no more than two lines for situations/hazards on the same link of that motorway. Alternating VMS should not be used (West European Road Directors (WERD) Deputy European Road Directors (DERD) 2000). Three-line VMS are common in the U.S., although two-frame messages (thus, at least four lines) are not rare.

12.3.2.3 Message Format

Reading time is significantly affected by the way information is placed and arranged on the VMS. According to Dudek (2002), the order of information depends upon

TABLE 12.2
Format Order for Text Messages

1. Format order when *incident descriptor* message is used for incidents	**2. Format order when *incident descriptor* message is used for roadwork**
Message elements for lane closure incidents[a]	**Message elements for lane closures[b]**
Incident descriptor	Roadwork descriptor
Incident location	Lane closure location
Lanes closed (blocked)	Lanes closed
Audience for action (if needed)	Audience for action (in needed)
Action	Action
Good reason for following action	Good reason for following action
3. Format order when *incident descriptor* message is replaced by or combined with the lane closed message element for incidents	**4. Format order when *incident descriptor* message is used for roadwork**
Message elements for lane closure incidents[c]	**Message elements for lane closures[d]**
Lanes closed (blocked)	Lanes closed
Lane closure (blockage) location	Lane closure location
Audience for action (if needed)	Audience for action (if needed)
Action	Action
Good reason for following action	Good reason for following action

[a] Message elements for freeway closure incidents are the same but not including the now obvious point 6.

[b] Elements for freeway closures are the same but now "closure location" is the element for point 2 and point 6 is eliminated.

[c] For freeway closure incidents substitute point 1 by "freeway closure (blocked)" and point 2 by "location of closure."

[d] For freeway closure incidents, substitute point 1 by "freeway closed" element and point 2 by "closure location."

whether an incident/roadwork descriptor message element is part of the message or whether the message is replaced by or combined with a "Lanes Closed" message element. Such a premise produces four basic VMS format order recommendations (see Table 12.2). Formats for single vs. two-framed VMS (distributing the elements in two frames) and for lane closure incidents vs. freeway closure incidents are also distinguished. Two-frame VMS are not unusual. A typical large text VMS can include three lines with 20 characters per line. In order to consider variations to the message elements due to changing conditions and the eventual need for reducing the amount of information displayed, a flow-chart VMS design process has been developed (Dudek, 2002; see Appendices B and C).

12.3.2.4 Bilingual Messages

Anttila et al. (2000) compared perception latency of single neutral messages (temperature) with two types of short, bilingual (Swedish, Finnish) messages (two words, one panel vs. two alternate Swedish–Finnish panels). About 81% of Finnish drivers understand some Swedish. Although presenting different information, bilingual messages took significantly more gazing time than monolingual ones. No differences were found between alternated and single presentation modes (displaying only one word and under low-flow conditions). Also, older drivers needed more time to fixate on the sign than younger subjects (Anttila et al., 2000).

Experiments performed by the Transport Research Laboratory with bilingual English–Welsh messages (the majority of English are unable to understand Welsh) indicate that the ratio of information recalled to information presented diminishes with bilingual messages: a 3-unit bilingual message was equivalent to a 12-unit monolingual message in terms of glance duration (MacLaverty et al., 1998a). Only highly specific, local, and worthy conditions (e.g., European–Muslims going through Spain toward Morocco on holidays) may consider bilingual use (SETRA, 1994). In sum, the use of bilingual signs should be limited to the most essential cases because such messages appear to heighten attention demands and visual distraction, especially with older drivers. This is especially marked when more complex messages are displayed or unfamiliar second languages are used (McLaverty et al., 1998a).

12.3.2.5 Abbreviations

The correct way of shortening words is language and culture specific. Abbreviations play a different role in European recommendations compared with the U.S. ones. In Europe, "abbreviations of words (except commonly known ones such as 'm' for meters, 'km' for kilometers, etc.), should be avoided because all drivers do not always understand them" (Mavrogeorgis et al., 1998; p. 26). In the U.S., abbreviations follow the MUTCD lists of (Dudek, 2002):

- Acceptable (e.g., HWY for highway).
- Acceptable but only with a prompt word preceding or following the abbreviation (e.g., HAZ — hazardous driving).
- Unacceptable.

Although a few abbreviations are recommended in Europe, the first two categories have about 90 terms in the U.S. In general, only clear and frequently used abbreviations, well known by the public, should be allowed. Improvising is not advisable; some short abbreviations well known by road engineers are totally exotic to drivers.

12.3.2.6 Message Absence

> Once a VMS system is installed, a question always arises concerning when messages should be displayed. There are two schools of thought: 1) display messages only when unusual conditions exist on the freeway; or 2) always display messages regardless of whether or not unusual conditions exist on the freeway. Or, as a minimum, always display a message during the peak periods and only when unusual conditions exist

> during the off-peak periods. The author of this report subscribes to the former of the two approaches because of human factors principles and because of difficulties in designing messages when incidents actually occur during the peak periods.
>
> **Dudek, C.L., 2002, pp. 3–7**

Hitchins et al. (2001, p. 1) explain the concerns. For example, when a VMS is blank, drivers cannot tell whether it works and therefore whether traffic conditions are normal. They also bring the counterargument to this point: VMS exhibiting unimportant information may cause unnecessary distraction to motorists. Also, perennial message exhibition may exhaust VMS' salient effect on drivers: "if signs are displaying a message of some sort all the time, drivers might not read the messages at all in time, on the basis that they are rarely important."

An interesting result emerges from drivers' opinions about blank messages when comparing English and French drivers (only 10 to 15% assume that the VMS was broken if there was no message) with Italian ones (75% would believe that a blank VMS is out of order). In fact, drivers' considerations emerge from respective policies, English and French adhering to the "blank" option and Italians to the "always display" one. If one is accustomed to seeing messages on a VMS, it is only normal to interpret a blank VMS as out of order. Interviews show that drivers like functional information, including some safety messages, whereas time, temperature, and courtesy messages are not liked as well (i.e., they are considered not so useful and rather annoying). In sum, "such messages should be employed with caution. All signs increase workload and could potentially be hazardous, in certain conditions. Additionally, the benefits of nontraffic messages have yet to be proven over the long term" (MacLaverty et al., 1998b, p. 49).

Hitchins et al. (2001) advocate for an interesting solution: to display familiar messages on selected VMS using a lower-case font (a way for drivers to categorize and distinguish nontraffic messages beforehand). It sounds fair, although such a solution restricts VMS to display lower case and leads to a "lights on all the time" situation. On the other hand, lower case has been recommended for informative traffic messages because they are faster to read (WERD/DERD, 2000), invalidating the former proposal. Some VMS incorporate flashes in order to attract extra attention to important messages. Now, political (to make a frequent, visible use of what citizens pay for) and economical (marketing of the information roads users want) arguments, not human factors and traffic ones, seem to push VMS toward a similar outcome: excessive VMS presence — and deliberate ignorance on the part of drivers. In the authors' view, more than being a creative solution, resorting to flashing lights could simply reflect an inadequate operational policy.

12.3.3 Pictogram and Combined Messages on VMS

Recommendations on pictograms and combined pictogram–text message design are common within areas such as Europe and Japan. Pictogram design, a very complex matter, especially when broad areas (national, international) are considered, will not be covered here (but is briefly mentioned at several other points in this book).

12.3.3.1 Two-Frame or Alternated VMS

A VMS with one pictogram on the left and three lines of text should suffice to convey an intended message. Using two frames here is problematic because many different elements may be used, thus complicating VMS interpretation. For example, when recurring to two-framed VMS, it is possible to repeat or combine two different pictograms. Even assuming that the primary pictogram was correctly chosen according to the particular road event, the selection of the second pictogram may introduce great variations. In addition, more room available for text may give way to redundancies (pictogram–text or text–text redundancies), making the VMS unnecessarily long. In sum, due to the differing quality of elements (pictograms, alphanumeric) and the possibilities open for combining them, two-framed pictogram–VMS should be rarely used and carefully controlled. In fact, only circumstances exceptional for high-speed roads (e.g., 60 km/h) allow such practice (SETRA, 1994; Arbaiza, 2000; Mavrogeorgis et al., 1998). This view contrasts with text-only VMS areas in which recommending two-framed formats is not uncommon.

12.3.4 Causes vs. Consequences: On VMS Reading Schemas

European VMS documents (Beccaria et al., 1991; Mavrogeorgis et al., 1998; WERD/DERD, 2000) insist on pictograms as the fundamental element on any VMS message, with minimized or short text. However, compared to regulatory and danger warning signs, informative and advisory signs frequently contain combined or text elements (as posted signs do).

Table 12.3 includes the basic message format recommended by WERD/DERD (2000) for regulatory, danger warning, and informative messages. Two basic types of informative messages can be distinguished: (1) *link* (referring to a hazard/event on the same road) and (2) *network* (concerning situations somewhere else on the motorway network or the network performance; restricted to some drivers), although the link message format basically follows the danger warning format. The adoption of a standard format for a third category, informative *rerouting* messages, is experiencing some difficulties due to the existence of many different VMS devices specifically for rerouting and, most importantly, due to the lack of a widely accepted and implemented pictogram for rerouting messages (Kenis, 2001).

In order to ease VMS comprehension, correct lexical elements must be determined, regular structure for presenting messages must be provided, and measures must be taken in order to avoid overloading. Again, the departure point is a standard combined VMS layout: one pictogram on the left plus two to three lines (12 to 18 characters per line). Expected (Western) reading schemas go from left to right and from top to bottom. Following the Vienna Convention, pictograms should represent the most important information on the message, with text as a supplement. Unlike text, pictograms can be understood by all types of drivers (immigrants, tourists, international drivers). This fact goes well in terms of structural processing features: things most important to communicate usually go first. When reading, first means left and the pictogram is found there. Regulatory, danger warning, and informative link messages accommodate that rule.

TABLE 12.3
Format Order for Regulatory and Danger Warning Messages

Basic format for regulatory messages

"Regulatory messages should be presented as signs, where necessary with matching additional subscriptions, both according to the Vienna Convention. Further additions to regulatory messages (symbols, texts) are optional, but not recommended. If used and therefore meaningful, they have an informative character" (p. 10).

Basic format for danger warning messages

"Messages concerning immediate danger warning should be presented as signs with matching subscriptions (both according to the Vienna Convention), or with appropriate new (yet to be developed and to be harmonized) signs/symbols. Further additions (symbols, texts) are optional, but should be readily understood" (p. 12).

Basic format

Pictogram area and/or line 1 (nature of the hazard) use a pictogram

Line 2 (distance to the hazard) and/or (extent of the hazard)

Line 3 (advice) or (additional information); note: part 3 is optional

Source: From WERD/DERD, Final version 3.0, Spring 2000. West European Road Directors (WERD), Deputy European Road Directors (DERD), p. 21, 2000.

However, in order to achieve a higher degree on standardization, it is necessary to define the general role of the pictogram. An early indication can be found in a document known as the *VMS White Book* (Beccaria et al., 1991, pp. 4–18): "If you tell drivers what they should do (whatever the combinations of signs and text), a reason should also be given, wherever possible." Similarly, recommendations within FIVE (framework for harmonized implementation of VMS in Europe) propose that informative link messages should preferably focus on consequences (e.g., waiting line, lane closed), not causes, as the primary information (see WERD/DERD, 2000). Drivers may like to know the reason why (i.e., the cause), but they must know (as specifically and accurately as possible) what they should do (i.e., the consequences for their driving). Drivers like relevant information for their driving and such information, more than causes, entails consequences. Holding to that rule, a general VMS element structure emerges: a pictogram always displaying consequences (on the left) and text complementing that main information on the right, normally on three rows of two to three lines each (that may or may not include the cause). Each line has a function/content assigned (see Table 12.3).

However, consider the example with a VMS with 18 characters per row in Figure 12.1. If one follows the main recommendations (Mavrogeorgis et al., 1998; WERD/DERD, 2000), the four examples are correct not only in terms of the number of rows (two) and the number of words (five, four) used, but also in terms of the recommended layout structure. The distance to event is placed on the second line; the first and the third messages place words ("Right Lane Narrows") in the first line (complementing the pictogram meaning); the second and fourth messages place

Consequences and causes

FIGURE 12.1 (See color insert following page 154.) Considering consequences vs. causes for national vs. international drivers.

words ("Road Work") in the third line (additional information concerning nature of event or causes).

Obviously, although local/national (here, English-speaking) drivers may have no problem decoding messages 1 and 3, international non-English speaking drivers will only make the most of consequence-oriented message number 4, because the pictogram indicates the opportune information in terms of direct actions concerning their driving. Finally, facing messages 1 and 3, English-speaking drivers still could see two different VMS for indicating the same event; such heterogeneity is undesirable. Here the display consequences principle allows for a determination of best practice. In sum, a principle of VMS design oriented to consequences seems adequate, regardless of the mother tongue, because such a principle contributes further to building a coherent logic on message design. Also, in this way, local, national, and international drivers might know or make a good guess about priority actions in terms of their driving.

12.4 CONCLUSIONS AND PRACTICAL IMPLICATIONS

Neither text-oriented VMS systems nor pictogram-oriented ones are without problems. The latter critically depend on an internationally valid pictogram set and on appropriate pictogram–text combinations. However, VMS need to "shrink" in order to suit the high-speed road context; this fact goes against text–VMS that need a minimum amount of words to inform (thus resorting to two-framed messages). With pictograms, more information can be obtained at a glance, although only when good, comprehensible pictograms are displayed. In both cases, harmonization of practices and standardization are essential, perennial tasks in order to avoid new road signs sharing the same problems identified for traditional road signs and adding further ones. These tasks are also necessary to avoid exhibiting highly salient electronic information in the wrong way, at the wrong place, and at the wrong moment because then this information competes inadequately with other driving priorities. In sum, new road sign systems base their promise of efficiency on a *better* use of *more* information, but for this "more is better" to be true, special care and attentiveness must be given to the real information reaching the road user.

ACKNOWLEDGMENTS

This research has been sponsored by the Spanish Road Directorate (Dirección General de Tráfico) within the frame of the ITS program jointly developed by the Spanish Road Directorate and the Traffic and Road Safety University Institute–INTRAS (University of Valencia). Commentaries and fruitful suggestions provided by Cándida Castro and Tim Horberry are gratefully acknowledged.

REFERENCES

Allen, T.M., Lunenfeld, H., and Alexander, G.J. (1971). Driver information needs. *Highway Res. Rec.*, 366, 102–115.

Anttila, V., Luoma, J., and Rämä, P. (2000). Visual demand of bilingual message signs displaying alternating text messages. *Transp. Res. Part F*, 3, 65–74.

Arbaiza, A. (2000). Manual de estilo de la señalización variable. En J.V. Colomer y A. García (Eds.): *Calidad e Innovación en los Transportes (Vol. III)*. Actas del IV Congreso de Ingeniería del Transporte, 1367–1374. Valencia: CIT 2000.

Beccaria, G., Bolelli, A., Wrathall, C.W., Rutley, K.S., Schneider, H.W., Balz, W., Friedrich, B., Ploss, G., Cremer, M., Putensen, K., Naso, P.G., and Schlüter, M. (1991). White book for Variable Message Signs application. Sobrero: The VAMOS Consortium.

Cooper, P.J. (1997). The relationship between speeding behavior (as measured by violation convictions) and crash involvement. *J. Saf. Res.*, 28, 83–95.

Dudek, C.L. (2002). Changeable Message Sign Operation and Messaging Handbook. (Draft). FHWA. U.S. DOT. http://tmcpfs.ops.fhwa.dot.gov.

Dudek, C.L. and Huchingson, R.D. (1986). Manual on real-time motorist information displays. Washington, FHWA/DOT.

Eco, U. (1995). *Tratado de Semiótica General.* Barcelona: Lumen.

European rules concerning road traffic, signs and signals (1968/1971). Vienna/Geneva: European Conference of Ministers of Transport.

Fabre, F., Klose, A., and Rathery, A. (1986). Electronic and traffic on major roads. Technical, regulatory and ergonomic aspects. Luxembourg: Commission of the European Communities.

Friedrich, B., Engels, A., Reischl, A., and Zhang, X. (1999). Tropic Results Catalogue. Traffic Optimization by the Integration of Information and Control — TROPIC deliverable D11.4. Birmingham, U.K.: W.S. Atkins Consultants.

Hitchins, D. Brown, T., McCoy, D., Quinton, M., O'Halloran, M., and Plewes, M. (2001). Lowercase font set development for Variable Message Signs (VMS). *Proc. 8th World Congr. ITS.* Sydney, 2001.

Kenis, E. (2001). Re-routing information issues for traffic management (including the rerouting sign). Traffic Management Plans European Workshop. Valencia: 5–6 June 2001.

Krampen, M. (1983). Icons of the road. *Semiotica*, 43(1/2), 1–203.

Lewis, B.N. and Cook, J.A. (1969). Towards a theory of telling. *Int. J. Man–Machine Stud.*, 1, 129–176.

MacLaverty, K., Buckle, G., Rämä, P., Luoma, J., Harjula, V., and Pauzie, A. (1998a). Information overload summary report. Traffic optimisation by the integration of information and control — TROPIC deliverable D12.3. Birmingham, U.K.: W.S. Atkins Consultants.

MacLaverty, K., Pauzie, A., Gaunt, G., Bruyas, M.P., Trauchessec, R., Deleurence, P., Ghuillon, V., and Le Breton, B. (1998b). Message absence: nontraffic messages displayed on VMS. Traffic optimisation by the integration of information and control — TROPIC deliverable D12.5. Birmingham, U.K.: W.S. Atkins Consultants.

MASTER (1999). Managing speeds of traffic on European roads. Luxembourg: European Commission.

Mavrogeorgis, T., Bonaldo, A., Faccio, G., Ferrante, E., Remeijn, H., Pauzie, A., Balz, W., MacLaverty, K., Luoma, J., and Rämä, P. (1998). Text and combined reference manual. Traffic optimisation by the integration of information and control — TROPIC deliverable D05.3. Birmingham, U.K.: W.S. Atkins Consultants.

Nenzi, R. (1997). Use of dynamic signing (VMS). Volume 3C. Telematics on the Trans European road Network 2 — TELTEN2. Final Report. Brussels: ERTICO.

Ranney, T. A. (1994). Models of driving behavior: a review of their evolution. *Accident Anal. Prev.*, 26, 733–750.

Riemersma, J.B.J., Moraal, J., and Godthelp, J. (1986). Human factors considerations relevant for highway information systems. In Fabre, F., Klose, A., and Rathery, A. (Eds.), Electronics and traffic on major roads. Technical, regulatory and ergonomic aspects, 137–144. Luxembourg: Commission of the European Communities.

Sagberg, F. (1999). Theoretical background. In Guarding Automobile Drivers through Guidance Education and Technology — GADGET: Visual Modification of the Road Environment. http://www.kfv.or.at/gadget.

SETRA (1994). Panneaux de signalisation à messages variables. Bagneux: Service d'Études Techniques des Routes et Autoroutes — SETRA.

TEMPO (2002). TEN-T workshop: Variable Message Signs: the users' experience. Valencia, October 28–29, 2002. ARTS, DGT.

WERD/DERD (2000). Framework for harmonized implementation of Variable Message Signs in Europe. Final version 3.0, spring 2000. West European Road Directors (WERD), Deputy European Road Directors (DERD).

Wogalter, M.S., Sojourner, R.J., and Brelsford, J.W. (1997). Comprehension and retention of safety pictorials. *Ergonomics*, 40(5), 531–542.

13 Some Critical Remarks on a New Traffic System: VMS Part II

Antonio Lucas and Luís Montoro

CONTENTS

13.1 Introduction 199
13.2 A New System 200
13.3 VMS: Expected Improvements 201
13.3.1 Old Problems 201
13.3.1.1 Posted Signs 201
13.3.1.2 Road Signs: Symbols and Icons 203
13.3.1.3 Road Signs: Three Perspectives on Sign Reading and Understanding 204
13.3.2 New Problems 206
13.3.2.1 Multiple Presentation Systems 206
13.3.2.2 Multiple VMS Content and Layout 206
13.3.3 Extending the Concept of (Variable) Road Signs 209
13.4 Conclusion and Practical Implications 210
Acknowledgments 210
References 211

13.1 INTRODUCTION

Information technologies are aiding the growth of new and more rational road transport systems. At the core of Intelligent Transport Systems (ITS), traffic management and control critically depend on technical devices and road information well suited for road users because, in the end, the information in front of road users (e.g., VMS) is the basic tool for improving road traffic. In addition to a necessary technological optimism, a critical view is necessary for lessening or avoiding pitfalls. New presentation systems may distort the road sign system and worsen communication to road users.

0-415-31086-5/04/$0.00+$1.50

13.2 A NEW SYSTEM

In 1900 approximately 100,000 cars were in the world. In 2000 this number had grown to 1000 million cars (Moustacchi and Payan, 1999; Organization for Economic Cooperation and Development (OECD), 1997). During this time, road infrastructures (also vehicles) have been permanently improved; there are more and better highways and motorways in terms of road signaling, geometry, pavement quality, visibility, etc. (OECD, 1990). In the late 1930s in the U.S. and in the early 1950s in Europe, access to private transport experienced a steady growth — with about 55 million cars in the world in 1950 — and so did the density of traffic flows, congestion, and traffic accidents. This density doubled in these two decades in most European countries, reaching some 91,000 fatalities in 1970 (Barjonet, 1997).

At this time and in spite of its continuing growth, limits to road infrastructure expansion were progressively foreseen. More roads mean more cars because, in road traffic, it "paves the way" to demand (Vester, 1997). More roads also mean considerable maintenance costs, pollution, noise, accidents, etc. The current principle is to optimize existing infrastructures by controlling and managing traffic flows (e.g., incident management, rerouting strategies) rather than to expand the number and capacity of roads. Accordingly, as Camus and Fortin (1995) explain, public road authorities are expanding their objectives from *traffic control*, which aims to influence the behavior of each user directly in the presence of a difficulty, to *traffic management*, which aims to shift or modify traffic flows, sometimes at a considerable distance from the problem area, in order to optimize the management of a road network as a whole.

The possibility of using advanced information and communication technology to enhance road traffic has been brought up on more than one occasion during the development of road transport. The Aigrain Report in 1969, for instance, suggested bringing electronic aids to traffic flows as an objective of cooperation between European countries, European Countries-Cooperation in the Field of Scientific and Technical Research, (EUCO-COST, 1983). Following the European Transport Safety Council (ETSC; 1999a), three moments can be distinguished. The first wave came in the technologically optimistic 1950s, and the second was in the 1970s when computer development was very intensive, although technical development was unsatisfactory or too expensive. The third telematics wave came in the 1980s, eventually fulfilling the expectations of vehicle, road, and traffic control experts. "Intensive research and development programs started simultaneously in Europe (Prometheus and Drive) and in Japan (AMITCS, RACS). While efforts in the United States started somewhat later, development is now very rapid" (ETSC, 1999a, p. 14; also see Tignor et al., 1999).

The use of radio broadcasting and VMS came out in the 1970s as the most appropriate solutions to communicate traffic information to road users (COST 30 bis, 1985; Dudek, 1991). VMS represent a significantly distinct step as they follow, complement, and expand the concept of fixed, traditional road signs within the road sign system. The idea behind major engineering programs such as Prometheus and Drive is to combine information technology with the road transport system to increase efficiency.

> These promises are continuously presented in policy documents from various government and private actors in the field. The problems of congestion, pollution and accidents in road transport are to be handled as information problems. By increasing the supply of information to road users, the road transport system is to become more harmonious and orderly.
>
> **Juhlin, O., 1997, p. 175**

In sum, massive, complex, and expensive information technology systems designed and implemented for improving the road transport system efficiency are based on two premises: (1) the problems in the road transport system can be envisioned as information problems and (2) increasing the supply of information to road users will make a better road transport system. Eventually, improvement in the road transport system is to be achieved by presenting the right information at the right place in the right moment in a short, clear, understandable way to road users.

13.3 VMS: EXPECTED IMPROVEMENTS

ITS effectiveness depends on the design of road signs presented to drivers, which in turn must deal with existing signing problems and difficulties. Although some problems can be surpassed by VMS use, others are new, brought by ITS implementation.

13.3.1 Old Problems

Two aspects of the road sign system are introduced: drivers' ignorance of fixed road signs and the internal characteristics of the road sign system.

13.3.1.1 Posted Signs

The road transport system is organized following the rational bureaucratic model: there are some written norms (highway code), and road users assume their roles (pedestrian, driver, police) and behave according to their positions within the system (priority norms, speed limits, etc.). ITS do not change those basic principles and must serve only to increase system organization and to provide better mobility distribution. Driving behavior is conceived in a similar way within fixed and variable road sign systems: drivers process road information and make decisions according to what road signs demand. As seen throughout this book, the relationship between road users and traffic signs is a complex one because road signs are objects that are seen and understood as well as norms to comply with.

VMS may help to overcome some problems observed with posted signs, notably, disregarding signs:

> Excessive sign posting can cause informational overload and lead to general ignoring of traffic signs by the drivers, so that even the crucial information is not perceived any more. In many countries there exists a tendency to compensate failures in road layout with overregulation through traffic signs. This overregulation mainly serves the interests of authorities in the case of accidents to avoid legal problems.
>
> **Schmotzer, C., 1999, p. 6**

Such factors create a feedback effect: the frequency and diversity of road signs discourage attention and may increment the potential for accidents.

> As most engineers and TLE [traffic law enforcement] professionals know, poor design generates more violations — either through mistakes, frustration, disrespect of what looks like stupid demands, or via other mechanisms. Removing design faults can dramatically change apparent compliance levels.
>
> **Zaidel, 2001, p. 23**

For example, not coming to a full stop may lose its relevance when appropriately replacing a YIELD sign by a STOP sign (an issue explored by Ray Fuller earlier in this book).

The extended view that posted signs refer to real traffic conditions only to some extent (e.g., speed limits) can be counterbalanced by VMS' displaying highly relevant, specific, and useful road information. That is why, in order to preserve its value of salience, some specialists consider that a VMS should be left blank when nothing important needs to be transmitted to drivers (see the previous chapter). The VMS system credibility may lessen when displayed information is inaccurate, not current, irrelevant, obvious, repetitive, trivial, or erroneous (Dudek, 2002). By the same token, specific messages demanding specific behaviors (position on the road, distance to event, speed, head-following distance) should be used instead of generic messages ("caution," "near"; Luoma and Rämä, 2001). Finland constitutes a good example of this "VMS philosophy" because it is the only European country in which the congestion pictogram is restricted to VMS use (Nenzi, 1997).

Speed regulation. Speed is a major cause of road accidents, and speed regulation is undoubtedly an opportune case for VMS use:

> There is no doubt that the present system for setting speed limits and enforcing their compliance is a colossal failure. I expect Europe is similar to America where practically everybody drives faster than the speed limit most of the time and enforcement is infrequent and sporadic. When looking at this massive failure, the first thing a behavioral scientist would ask is, "Who's setting the speed limit? Why do 90% of the people think it's perfectly safe to go faster? Are we all so stupid that we are taking unnecessary risks? I think a strong argument can be made that the speed limits are set too low and that we should consider more realistic, variable speed limits. In such a system, the speed limit on a given road would be set at a level that changed over time depending on the prevailing conditions of the highway, adverse weather and traffic congestion.
>
> **Bower, G.H., 1991, p. 8**

Recently, the British administration has also recognized that fact: for accidents related to speed to be reduced, particular care is needed in order to avoid the "inconsistency in the way speed limits have been set throughout the country" (Department of the Environment, Transport and Regions (DETR), 2000, p. 22). Adjusting speed limits to road conditions may contribute to improving compliance to speed limits and to reducing accidents (Bolte, 2002). The assumption is that drivers will comply with variable speed limits as long as the speed limit imposed is more or less in accordance with their judgments of the right speed for the road section at that particular moment. VMS design has little to do with compliance here as long as an official pictogram is displayed (and understood).

However, other speed regulation systems and procedures have been explored in order to achieve compliance with speed limits more akin to VMS *design strategies*, for example, the studies carried on by Van Houten and collaborators on the effect of posted exhibition of speed compliance upon real speed and accidents (Van Houten and Nau, 1981; Van Houten et al., 1985; Rothengatter, 1988). In short, exhibiting the percentage of drivers complying with the speed limit induces more drivers to comply with (nearby) posted signs and/or to reduce their speed in a particular road section. Interestingly, Groeger and Chapman (1997) found similar effects with two traffic violations, excessive speed and headway distance. In a series of experiments carried on with a driver simulator, they obtained a violation reduction by showing VMS displaying 80% compliance rates, but only when tested drivers observed that other drivers were also complying with regulations (i.e., leaving more headway distance or reducing speed).

13.3.1.2 Road Signs: Symbols and Icons

Krampen (1983) carried on a historical sociosemiotic study of the road sign system following the classical typology of signs introduced by the American philosopher Charles S. Peirce. Basically, *icons* acquire their meaning because they are similar to what they portray, while the meaning of *symbols* is a matter of social convention. According to Krampen, a number of parameters allow one to establish the evolving trends of the road sign system: increasing number of road signs and road sign categories (differentiation); a trend from "conventionalizing" toward "iconizing" of road signs (forced by internationalization and the need to overcome language barriers); and focusing on sign pictures, a trend from more realistic to more abstract designs.

Although symbols and icons involve some degree of learning, icon learning is more straightforward because it is ascribed to a logical code, well learned within the general culture. These combined trends situate pictograms as the main protagonists of the road sign system. Road signs change as a result of historical factors, both external (introduction of new technology, increasing speed, emerging pressure of automobile and touring clubs) and internal (shapes, colors, and pictorials determining the correspondences at the level of content expression within the system). The growth of the official road sign system takes place in a process in which the internal linguistic possibilities and limitations interact with external social and technological determinants (see Section 13.3.2.2).

Prosymbolic and proiconic systems. The dominance of iconic or symbolic signs is clearly established in the three main ITS world areas: Japan, North America, and Europe. Although all three areas use and mix text, conventional pictograms, and iconic pictograms, Japanese (see http://www.its.go.jp/ITS) and European VMS systems are more icon oriented, whereas the U.S. VMS system is text (symbol) oriented (see Chapter 12; Tignor et al., 1999).

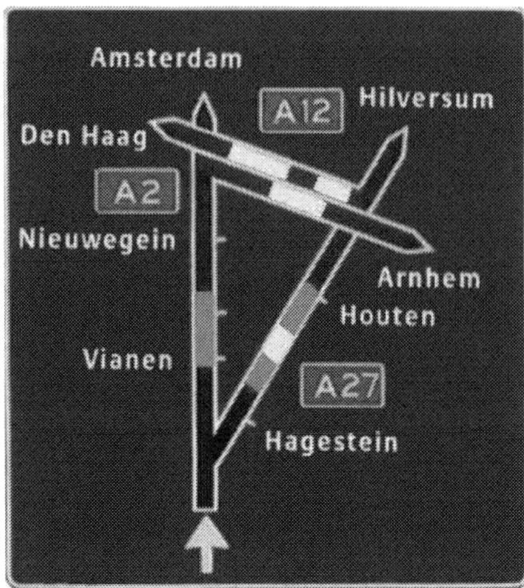

FIGURE 13.1 (See color insert following page 154.) Real-time graphic displays showing information on congestion and travel time. Left: traffic mimicry panels in M-40 (Madrid); right: real-time graphic display (Dutch prototype).

Accordingly, VMS design problems in Europe come mainly from pictogram selection and comprehension, combined pictogram–text strategies, and message standardization, whereas in the U.S. problems come from excessive numbers of message units, abbreviations, and two-frame message use (also from non-English readers). Japan leads the use of graphic representations of motorway maps, for example, showing congestion levels within certain sections on the road network by using colors. Some European and North American administrations are now initiating the use of real-time graphic displays *a la Japanese* (see Figure 13.1). Although pictogram use has been considered, research carried out on American drivers' comprehension of some versions of pictograms currently used in Europe (e.g., accident, congestion) has not been satisfactory. Also, infrastructure problems (e.g., ordinary VMS graphical display capabilities) exist. "Until highway agencies can afford to install stadium and arena type full-matrix, full-color signs, use of graphics and symbols will be limited" (Dudek, 2002, pp. 5–41).

13.3.1.3 Road Signs: Three Perspectives on Sign Reading and Understanding

Modern traffic signs can be roughly located in an icon–symbol/pictogram–text continuum. The most iconic road signs try to represent reality as such, for example, graphic representations of motorway maps. Then come pictograms that may be mainly iconic (e.g., road works) or mainly symbolic (e.g., speed limit). Finally, road signs based on text are utterly conventional. This continuum should bear an influence on the ideas about sign interpretation processes on the part of drivers.

Juhlin (1997) distinguishes three approaches for road sign interpretation:

- In the *algorithmic approach*, the sign generates the interpretation on the part of road users. This approach coincides with the interpretation of "pure" icons.
- In the *routine approach*, road users learn the meanings of road signs, normally through formal education. They memorize their meanings (convention) and then practice reinforces the correspondence meaning–sign (symbol).
- In the *situated approach*, drivers learn to interpret road signs according to what other drivers do. Followed social practice and action on the road gives road signs their meaning, so the sign system cannot be changed without also changing forms of life and practices: "... the information systems should be as open as possible for the users to create interpretations and messages as resources in their practice. The messages should support rather than steer the actors' dialogues" (Juhlin, 1997, p. 192).

Juhlin's observations run parallel to social considerations on road sign design to increase compliance. To some extent, such considerations may be bypassed here because VMS are devices located in high-speed roads where little interaction is expected between drivers. Symbolic interation concepts can be borrowed from sociologist E. Goffman (1959) to describe, for example, drivers' passing in a motorway as *encounters* (vague mutual awareness) rather than as *performances* (mutual recognition) and to locate VMS more in a context for encounters. Therefore, interpretation of current VMS has more to do with the algorithmic and routine approaches than with the situated one.

Returning to Krampen's (1983) analysis and to the extent that the road sign system should enjoy universal intelligibility, icon–pictograms should be the basic element: "... they speak by themselves." However, it seems that pure universal icons for depicting traffic situations are difficult to find; the exactly right situation must be determined and then depicted in a simple way so that it can be represented with a simple set of graphic resources (lines, colors) within a limited space with a given shape (triangles, circles, etc.). Basic perception and comprehension problems easily arise. The problem extends further because road sign systems need to grow quickly due to the complexity of the road transport system and the increasing number of traffic situations to be regulated.

Finding a good pictogram is not easy. In the middle 1980s seven new pictograms were proposed for study and VMS use in the frame of EUCO-COST (1983). COST-30 bis (1985) took five of these seven as suitable recommendations; 20 years later, only two pictograms (congestion, accident) survive (Hubert et al., 1998), and improved pictograms are still needed for very important daily events such as fog, snow, oncoming vehicle, restricted lane, and diversion. (Luoma and Rämä, 2001).

Such considerations have important practical implications. If new signs are introduced into the respective national highway codes, new drivers will learn them, but what happens with drivers already present in the system? Icon–pictograms may give a better chance for inference and interpretation. Pictograms based on convention depend on learning and memory, which may be a small problem when dealing with presumably slow-growing traditional posted metal road signs for which design is selected and implemented once and forever. However, it can cause trouble precisely with spreading, flexible, and adaptable road signs such as VMS.

13.3.2 New Problems

13.3.2.1 Multiple Presentation Systems

ITS designers want their systems to show messages that drivers may interpret in a nonambiguous way, do not perturb the driving task, are universal, go beyond drivers' competence and intentions, and are "transparent" (Juhlin, 1997). Presentation systems are essential because they represent the final informational link with the driver. As the next chapter in this book explains, before the ITS "revolution," posted signs were the only concrete presentation system representing a unique road sign system as found in any highway code. Now, many potential presentation systems (e.g., VMS) may, within or outside the car, show pictograms, maps, texts, sounds, etc., thus extending the road sign system as it appears to drivers.

13.3.2.2 Multiple VMS Content and Layout

Traditional posted signs are always in the same place, adopt approximately the same size, and incorporate the same elements (pictorial, alphanumeric) distributed in the same way in the plate. Posted signs are designed once; conversely, VMS devices may be switched on and off, show different messages, and show the same basic messages in different ways: different number and type of elements (text, pictogram, combined); format; structure; and distribution on the panel. In addition, VMS may be designed on a daily basis by different sign producers (e.g., traffic management center operators). The need exists for determining the purpose of the message (message categorization), and that step may introduce heterogeneity because different operators may consider different salient features as the relevant ones for defining the same situation. Even if the basic message parameters are correct (see Figure 13.2), however, heterogeneity may surge because different pictograms, words, and distribution of elements over the panel (optional or imposed by the specific technical design on a particular VMS) are possible.

- *The technical problem: display capability.* VMS displays (elements, shapes, colors, formats) depend on technological possibilities. Evolution has not stopped with VMS technology (from rotating prisms to incandescent lamps to optic fiber to Light Electric Diode (LED)) and communication (from copper to optic fiber) (Dudek, 1991; Arbaiza, 2000). Some problems remain (glare, deterioration, available colors) but, especially after the LED revolution, even the most complex signs can be displayed on VMS. However, in the middle 1980s measures such as color inversion (using black instead of white) were suggested to adjust pictogram display to technical possibilities (COST-30 bis, 1985).
- *The innovation problem: on new signs' design.* The innovation problem refers to new traffic signs' design. For example, the COST 30 experts

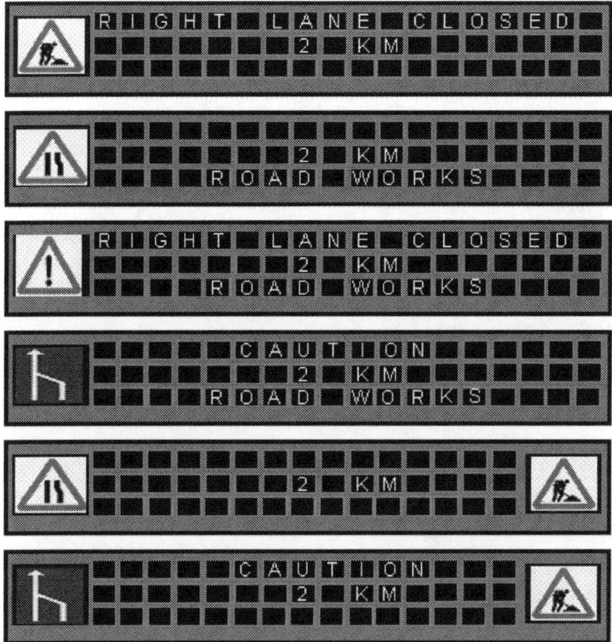

FIGURE 13.2 (See color insert following page 154.) Examples of different "correct" VMS alternatives for indicating "Caution. The right lane is closed in 2 km due to road work." Pictograms are in agreement with 1968 Vienna Convention; formats are in agreement with WERD/DERD (2000).

assumed that Vienna Convention signs would continue to be used on VMS when appropriate, but the introduction of high-speed roads coupled with greatly improved vehicle performance created situations unforeseen at the time of the convention (EUCO-COST 30, 1985). Very important VMS events, such as "congestion," "accident," or "slippery road," did not have a corresponding pictogram at the moment. Some of these problems are still unsolved (see Figure 13.3).

- *The administrative problem.* The administrative problem refers to the slowness and complexity of the process of changing or amending international accords. Standards adopted in the Vienna Convention for fixed signs were implemented only 10 years later. Amendment proposals incorporating VMS issues (color inversion, lane regulation signs) date from the 1980s but did not enter the convention until the middle 1990s (Nenzi, 1997). Many cultural, political, and technical factors must be taken into account and enter the process; many studies need to be carried on to decide the international adequacy of changes or new traffic signs.

  Road sign systems, built to last, must be coherent and stable to be learned and used by millions of drivers. However, many countries individually adapt VMS design (new signs and/or layout) in order to solve problems that may have priority at the local or national level; these signs have no

Text only VMS

Pictogram VMS

Combined pictogram-text VMS

Combined pictogram-text VMS

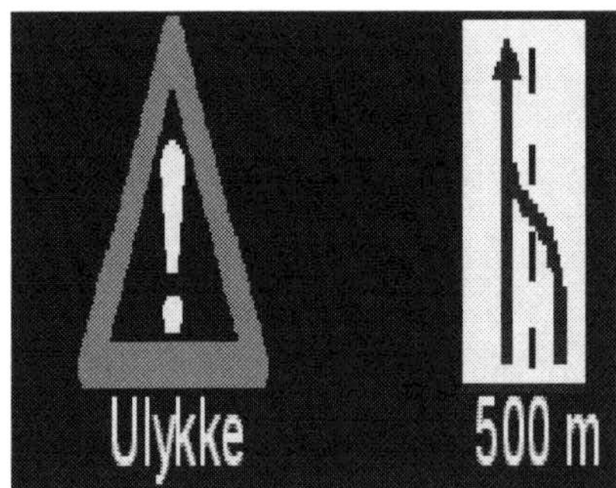

Combined (vertical layout)

Combined (horizontal layout)

FIGURE 13.3 (See color insert following page 154.) Examples of VMSs with different layout, structure for informational elements, and disposition on the panel.

international status. On the one hand, this is positive because some designs are trial answers to local matters and because (with some empirical control) this constitutes a useful possibility for innovative solutions to problems that are probably experienced by other countries (e.g., VMS sign and infrastructure solutions for indicating diversion, journey times, congestion, etc.). On the other hand, the proliferation of new signs makes it more difficult to achieve standardization and harmonization of VMS practices (Meekums et al., 2002).

- *The technological problem.* When analyzing the development and implementation of VMS in Europe, COST programs pointed to two fundamental and related problems: the need for fixing symbol standards (i.e., Vienna Convention) and the need for leaving margin to foreseeable technical development. Regarding the latter, the decision was that "standards for manufacture would not be necessary, the adoption of functional standards (signs) would be sufficient. ... The equipment used for displaying messages on changeable message signs should not be standardized internationally" (COST-30 bis, 1985, pp. 38–41).
  An early specification of technical standards would have been negative for VMS technical improvement. Indeed, innovation has been beneficial and in the last years perception and display capabilities have been greatly improved. On the one hand, however, many road administrations must now confront the problem of VMS device heterogeneity; on the other hand, technical improvements and innovations may invade the area of Vienna Convention standards, threatening standardization (signs, colors, and shapes; layout; drivers' reading patterns) and producing heterogeneity on displayed messages (Figure 13.3).

In sum, VMS may be improved by technology (i.e., easier to perceive and representing more sophisticated signs), to the benefit of all drivers. However, unrestricted technological innovations introduce difficulties in terms of integrating and harmonizing the whole VMS system (infrastructure, signs, and layout heterogeneity; greater message variability), which is detrimental to drivers and to safety and efficiency of road sign systems as a whole.

13.3.3 Extending the Concept of (Variable) Road Signs

Other drivers play a role in understanding of and compliance with road signs, as well as in the way that driving context (Rothengatter, 1991; Juhlin, 1999) is interpreted. Although driving is acknowledged as a social and interactive activity in which communication and cooperation are essential (Wilde, 1976; Zaidel, 1992), the traffic sign system has been designed for individual (i.e., collective but isolated) obedience. For example, cars — some exhibiting commercial information — are very poor traffic sign conveyors, with little more than horn, lights, and blinkers.

The idea about information flowing on the road could expand, however. For example, consider simultaneous use of the four car blinkers, originally designed and implemented on vehicles for drivers to indicate an atypical stop on the road (normally on hard shoulders) due to emergency or malfunction.* More and more drivers use this, particularly when approaching an anomalous line on motorways (fast speed) in order to avoid rear collisions, i.e., as an *unofficial* road sign warning of danger. In fact, this particular action of drivers coincides with several studies indicating that rear collisions are very frequent events (ETSC, 1999b). Compared to other measures (warning congestion on posted signs, or even VMS), this naturally adopted social measure is highly specific: it may be switched on at any time by any driver and last only a few seconds. Current traffic management centers can rarely convey this highly subtle danger warning sign.

13.4 CONCLUSION AND PRACTICAL IMPLICATIONS

Official and unofficial road signs are currently undergoing promising research and professional and policy inquiries, hopefully to aid mobility and road safety. It is clear, though, that ITS may promote a heterogeneous, uncontrolled extension of the road sign system, thus making interpretation on the part of road users more difficult. For example, consider basic VMS textual messages recommended in the U.S. or the ALERT-C catalogue sum 280 and 300 messages, respectively (Dudek, 2002; Juhlin, 1997). New pictograms are demanded for old and new road situations; these designs are sometimes developed in an uncontrolled and local manner (WERD/DERD, 2000; Trans-European Intelligent Transport Systems Project (TEMPO), 2002).

In addition to changing road information elements (e.g., pictograms, abbreviations, and verbal labels), new VMS device structures force the use of different message formats, making road sign harmonization and coherence all the more difficult. Clearly, for road sign systems and road user interpretation to be preserved, innovative practices should be encompassed by standardization requirements on VMS design practices (see the previous chapter). Sign dissemination in the road system must consider the specific relationship among sign, task, and user in terms of comprehension and interpretation processes.

ACKNOWLEDGMENTS

This research has been sponsored by the Spanish Road Directorate (Dirección General de Tráfico) within the frame of the ITS program jointly developed by the Spanish Road Directorate and the Traffic and Road Safety University Institute — INTRAS (University of Valencia). The images in Figure 13.1 and Figure 13.3 are courtesy of Alberto Arbaiza; Hans Remeijn and Bas Schenk; Eugene O'Connor; Robert Ridgway; Kenneth Kjemtrup; and Jean Marc Chauvin from the Spanish, Dutch, Irish, English, Danish, and French road administrations, respectively. Commentaries and suggestions provided by Cándida Castro and Tim Horberry are gratefully acknowledged.

* We would like to thank Enrique Reoyo (INTRAS) for his valuable comments on this issue.

REFERENCES

Arbaiza, A. (2000). Manual de estilo de la señalización variable. In J.V. Colomer and A. García (Eds.), *Calidad e Innovación en los Transportes (Vol. III). Actas del IV Congreso de Ingeniería del Transporte*, 1367–1374. Valencia: CIT 2000.

Barjonet, P. (1997). Transport psychology in Europe: a general overview. In T. Rothengatter and E. Carbonell: *Traffic and Transport Psychology. Theory and Application*, 21–30. Oxford: Pergamon Press.

Bolte, F. (2002). Variable Message Signs in Germany. Experiences, Guidelines and Strategies. VIKING Domain 3 workshop on VMS. 5–6 November, 2002. Gothenburg.

Bower, G.H. (1991). Incentive programs for promoting safer driving. In M.J. Koornstra and J. Christensen (Eds.): *Enforcement and Rewarding: Strategies and Effects*, 8–18. Leidschendam: SWOV.

Camus, J.P. and Fortin, M. (1995). Road Transport Informatics. Institutional and Legal Issues. Study drawn for European Conference of Ministers of Transport. ECMT and ERTICO. Paris: ECMT.

COST 30 bis (1985). Electronic traffic aids on major roads. Luxembourg: Commission of the European Communities.

DETR (2000). Road traffic penalties. A consultation paper. Home Office Communication Directorate. U.K. http://www.homeoffice.gov.uk.

Dudek, C.L. (1991). *Guidelines on the Use of Changeable Message Signs*. Washington: FHWA.

Dudek, C.L. (2002). *Changeable Message Sign Operation and Messaging Handbook*. (Draft version). FHWA U.S. DOT. http://tmcpfs.ops.fhwa.dot.gov.

ETSC (1999a). *Intelligent Transportation Systems and Road Safety*. Brussels: ETSC.

ETSC (1999b). *Police Enforcement Strategies to Reduce Traffic Casualties in Europe*. Brussels: ETSC.

EUCO-COST 30 (1983). European project on electronic traffic aids on major roads. Final Report (EUR 7154). Luxembourg: Commission of the European Communities.

European rules concerning road traffic, signs and signals (1968/1971). Vienna/Geneva: European Conference of Ministers of Transport.

Goffman, E. (1959/1990). *The Presentation of Self in Everyday Life*. London: Penguin.

Groeger, J.A. and Chapman, P.R. (1997). Normative influences on decisions to offend. *Appl. Psychol.: Int. Rev.*, 46, 265–285.

Hubert, R., Remeijn, H., Rämä, P., Luoma, J., MacLaverty, K., Duncan, B., and Carta, V. (1998). Pictogram presentation and recommendations. Traffic optimisation by the integration of information and control — TROPIC deliverable D04.2/DO4.3. Birmingham, U.K.: W.S. Atkins Consultants.

Juhlin, O. (1997). Reflecting on road signs in an age of information technology. In *Prometheus at the Wheel — Representations of Road Transport Informatics*, thesis. Linköping: Linköping Studies in Arts and Sciences.

Juhlin, O. (1999). Traffic behavior as social interaction — implications for the design of artificial drivers. *Proc. 6th World Congr. ITS*. Toronto, 1999.

Krampen, M. (1983). Icons of the road. *Semiotica*, 43(1/2), 1–203.

Luoma, J. and Rämä, P. (2001). Comprehension of pictograms for Variable Message Signs. *Traffic Eng. Control*, 2, 53–58.

Meekums, R., Porooshasp, K., and Ridgway, R. (2002). European variable message sign harmonization issues. *Proc. 9th World Congr. ITS*. Chicago 2002.

Moustacchi, A. and Payan, J.J. (1999). *L'automobile: Avenir d'une Centenaire*. Paris: Flammarion.

Nenzi, R. (1997). Use of dynamic signing (VMS). Vol. 3C. Telematics on the trans European road network 2 — TELTEN2. Final report. Brussels: ERTICO.

OECD (1990). Behavioral Adaptations to Changes in the Road Transport System. Paris: OECD.

OECD (1997). Road Transport Research OUTLOOK 2000. Paris: OECD.

Rothengatter, T. (1988). Risk and the absence of pleasure: a motivational approach to modeling road user behavior. *Ergonomics*, 31, 599–607.

Rothengatter, T. (1991). Normative behavior is unattractive if it is abnormal: relationships between norms, attitudes and traffic law. In M.J. Koornstra and J. Christensen (Eds.): *Enforcement and Rewarding: Strategies and Effects*, 91–93. Leidschendam, SWOV.

Schmotzer, C. (1999). Road markings and traffic signs. In guarding automobile drivers through guidance education and technology — GADGET: visual modification of the road environment. http://www.kfv.or.at/gadget.

TEMPO (2002). TEN-T Workshop: Variable Message Signs: the users' experience. Valencia, 28–29 October 2002. ARTS, DGT.

Tignor, S.C., Brown, L.L., Butner, J.L., Cunard, R., Davis, S.C., Hawkins, H.G., Fischer, E.L., Kehrli, M.R., Rusch, P.F., and Wainwright, W.S. (1999). *Innovative Traffic Control Technology and Practice in Europe*. Washington, D.C.: FHWA/DOT.

Van Houten, R. and Nau, P.A. (1981). A comparison of the effects of posted feedback and increased police surveillance on highway speeding. *J. Appl. Beh. Anal.*, 14, 261–271.

Van Houten, R., Rolider, A., Nau, P.A., Friedman, R., Becker, M., Chalodovsky, I., and Scherer, M. (1985). Large-scale reductions in speeding and accidents in Canada and Israel: a behavioral ecological perspective. *J. Appl. Beh. Anal.*, 18, 87–93.

Vester, F. (1997). *El Futuro del Tráfico*. Madrid: Flor del Viento Ediciones.

WERD/DERD (2000). Framework for harmonized implementation of Variable Message Signs in Europe. Final version 3.0 Spring 2000. West European Road Directors (WERD), Deputy European Road Directors (DERD).

Wilde, G.J.S. (1976). Social interaction patterns in driver behavior: an introductory review. *Hum. Factors*, 18, 477–492.

Zaidel, D.M. (1992). A modeling perspective on the culture of driving. *Accident Anal. Prev.*, 24, 585–597.

Zaidel, D.M. (2001). Noncompliance and accidents. Working paper 3 for Project "ESCAPE" (enhanced safety coming from appropriate police enforcement). WP2 noncompliance and accidents. Contract No: RO-98-RS.3047. VTT, Finland.

14 A Sign of the Future I: Intelligent Transport Systems

Michael A. Regan

CONTENTS

14.1 Introduction 213
14.2 Traffic Signs: Here and Now 213
14.3 Intelligent Transport Systems 214
14.4 Signs on the Move 215
14.5 Traffic Signs inside the Vehicle 216
 14.5.1 Regulatory Information 216
 14.5.2 Warning Information 218
  14.5.2.1 Traffic Information Systems 219
  14.5.2.2 Advanced Driver Assistance Systems 220
 14.5.3 Guide Information 221
14.6 Conclusions and Implications 223
References 224

14.1 INTRODUCTION

This chapter reviews emerging technologies capable of presenting new and existing traffic information to road transport users in radically different ways from that of the traditional traffic sign. The human factors implication of these emerging capabilities are considered in the next chapter, primarily from the perspective of the road vehicle driver. The technologies and human factors issues discussed in both chapters, however, are relevant to the presentation of traffic information in other transport domains, such as rail and aviation.

14.2 TRAFFIC SIGNS: HERE AND NOW

The traffic sign is one form of traffic control device. As noted in several places in this book, the purpose of such devices is to aid in ensuring the safe, predictable, efficient, and orderly movement of traffic. Road traffic signs may be broadly

0-415-31086-5/04/$0.00+$1.50

classified into three groups: (1) regulatory signs; (2) warning signs; and (3) guide signs. As noted by Lay (Chapter 3), most signs must be designed so that they can, in a single visual glance, be detected, read, understood, and acted on in the intended manner.

Several characteristics of traditional road traffic signs can be discerned:

- Information presented on them is largely static and unchanging.
- Information on them is conveyed to the driver by a display (i.e., a sign) outside the vehicle.
- Information on them is conveyed visually to road users.
- All road users are exposed to the same traffic signs (although their interpretations of what they see may differ).
- Information on traffic signs is presented in the same way for all road users, even though they vary widely in their ability to detect, comprehend, and act on it.

Emerging traffic technologies, known collectively as "Intelligent Transport Systems" (ITS), are emerging that are capable of presenting traffic information to road users in a manner quite different from that of the traditional traffic sign. These technologies can provide road users with changing traffic information based on real-time traffic conditions. In addition, they can provide it to the driver from both outside and inside the vehicle via internal displays.

This chapter commences with a review of these developments, focusing first on the variable message sign (VMS). Technologies capable of presenting traffic information to drivers inside the vehicle are then discussed. The discussion is structured around the three main categories of information conveyed by traffic signs — regulatory, warning, and guidance information — and the manner in which emerging ITS technologies make it possible to display this information inside the vehicle. Additional information that can be conveyed to road users via these technologies is identified.

In the next chapter, the human factors implication of these technological developments are discussed. That chapter concludes with a consideration of whether, in light of these developments, it will be necessary in the future to have traditional road signs.

We begin by defining what is meant by the term *Intelligent Transport System.*

14.3 INTELLIGENT TRANSPORT SYSTEMS

ITS is an umbrella term for a collection of electronic, computing, and communication technologies that can be combined in various ways to increase the safety and mobility of the transport system and to reduce harm to the environment (Regan et al., 2001). Although most ITS technologies have not been deployed long enough to know how effective they are in enhancing safety, it is predicted that the safety benefits deriving from them will be enormous (Rumar et al., 1999). This has prompted many jurisdictions around the world to support the early deployment of those technologies deemed to have greatest safety potential. Three broad categories of ITS can be discerned:

- *Vehicle-based* ITS technologies consist of sensors on the vehicle (e.g., radar, global positioning system) that collect traffic data, onboard units (OBUs) that receive and process these data, and display units that issue messages and warnings to the driver within the vehicle. Following distance warning systems, for example, utilize forward-looking radar to determine if the host vehicle is following a vehicle ahead too closely and warn the driver if this is so.
- *Infrastructure-based* ITS technologies consist of roadside sensors that collect traffic data that are processed on site or remotely and then transmitted to the driver via roadside equipment such as a VMS. The advantage of these systems over vehicle-based systems is that traffic information and warnings derived from them are available to all drivers. In addition, they are able to collect traffic data that cannot be collected by vehicle-based systems, such as the presence of fog on the road ahead.
- *Cooperative-based* ITS technologies derive traffic information from the road infrastructure, from other vehicles on the road network, or from both sources and transmit this to the driver via VMS or via displays within the vehicle. Infrastructure-based ITS technologies, for example, can be used to detect a vehicle approaching an intersection and send a warning to other vehicles approaching the intersection of the presence of the first vehicle. Alternatively, vehicle-based ITS technologies in one vehicle can be used to warn another vehicle equipped with ITS technologies of its presence on the approach to an intersection, without any support from infrastructure-based systems.

The following sections describe emerging ITS technologies falling under one or more of these general categories that are capable of presenting traffic information to road users in quite different ways from that of the traditional road traffic sign.

14.4 SIGNS ON THE MOVE

Traditionally, information displayed on traffic signs has been static and unchanging; once painted onto the sign, the information remained there until the sign was replaced, deteriorated, or was removed. Traffic, however, is a dynamic process and conventional, static, road signs are not capable of reflecting changing traffic conditions.

As mentioned in the previous two chapters, recent developments in traffic technology have led to the emergence of so-called variable (or changeable) message signs. The VMS is a flexible interface in that it can be used to display regulatory, warning, and/or guide information. Given that it conveys messages that vary over time, the VMS shares properties of the traffic sign and the traffic signal. A special category of interactive VMS, the so-called vehicle-activated sign (VAS) (Winnett and Wheeler, 2002), has been designed to deliver targeted warning information to individual drivers when they exceed a particular vehicle performance threshold (usually related to speed or following distance). Road sensors, such as buried inductive loops or microwave detector heads mounted on top of signs, have been used, for example, to monitor individual vehicle speeds on the approach to bends or junctions. Vehicles exceeding a preset speed threshold cause the sign to illuminate and issue an external message to the driver such as "slow down." Other systems have been designed to advise the driver to increase his or her following distance if detected traveling too close to the car in front. These interactive systems appear to be effective in reducing vehicle speed, have low initial and recurrent costs, and are self-enforcing with high and prolonged compliance (Winnett and Wheeler, 2002).

In summary, the VMS is an example of an infrastructure-based ITS technology that, unlike the traditional traffic sign, is capable of presenting changing traffic information in a variety of display formats to road users. It incorporates elements of the traditional traffic sign and the traffic signal. Unlike the traditional road sign, in which one-way communication takes place between the sign and the road user, the VMS provides for two-way communication: the behavior of the driver can influence what is displayed on the VMS.

14.5 TRAFFIC SIGNS INSIDE THE VEHICLE

The fact that most traffic information has traditionally resided on road signs external to the vehicle is a matter of historical consequence, until recently, static road signs and VMS have been the main technological platforms available for conveying traffic information to the driver. ITS technologies, however, make it possible to duplicate on displays within the vehicle some or all of the information displayed on traffic signs and VMS outside the vehicle. New categories of traffic information that cannot currently be conveyed on static signs or VMS can also be displayed by these technologies inside the vehicle. These new technologies make it possible to sequence and code the information in flexible ways to facilitate information transfer.

This section reviews emerging vehicle-based and cooperative ITS technologies capable of displaying traffic-related information to the driver within the vehicle. As noted earlier, the discussion will be structured around the three general categories of traffic information currently displayed to road users: regulatory, warning, and guidance.

14.5.1 Regulatory Information

Regulatory signs inform road users of traffic laws or regulations that would be an offense to disregard (Standards Australia AS 1742.1, 1991). The various regulatory signs typically fall into a number of subcategories, which vary slightly from country to country:

- Movement — e.g., "Stop," "Give Way," "Roundabout".
- Direction — e.g., "One Way," "Keep Left," "No Entry".
- Pedestrian — e.g., "Children Crossing," "Pedestrian Crossing".

- Speed — e.g., "60," "Road Work," "School Zone".
- Parking — e.g., "1/2P," "No Standing," "Loading Zone," "Clearway".
- Miscellaneous — e.g., "No Overtaking or Passing," "Stop Here on Red Signal".
- Exclusive lane use — e.g., "Bus Lane," "Bicycle Lane".
- Bicycle/pedestrian — e.g., "Shared Footway," "Bicycle Path".

Technologies exist that make it possible to present information that currently resides on regulatory signs outside the vehicle to the driver via a display in the vehicle. A good example of an ITS technology capable of doing so is intelligent speed adaptation (ISA). This system, variants of which exist as advanced prototypes, can be configured as a vehicle-based system or as a cooperative system. The aim of ISA is to make drivers adhere more closely to the posted speed limit, and systems vary according to the level of intervention they provide in doing so. Most systems are vehicle-based and employ a global positioning system (GPS) linked to an on-board computer and a digital road map on which speed-limited zones are overlaid. Alternatively, or in addition, the local speed limit can be communicated cooperatively to a receiver unit in the vehicle via external transmitters located on or near speed limit signs.

The more passive ISA systems simply display a visual duplicate of the posted speed limit continuously to the driver; these are known as "advisory" systems (Carsten and Tate, 2000). Some advisory systems issue, in addition to or instead of the continuous display, a warning (visual, auditory, and/or tactile) if the speed limit or some predefined speed threshold is exceeded (e.g., Regan et al., 2002). Another class of ISA, the so-called "mandatory" variant, prevents the driver from exceeding the posted speed limit. Here, the system is linked to some or all of the control elements of the vehicle (e.g., throttle, brakes).

ISA systems can also be categorized according to the integrity of the information gathered and processed. So-called "fixed" systems simply inform the driver of the posted speed limit and/or warn him if it has been exceeded. "Variable" systems additionally inform the driver of locations on the road network where a lower speed limit applies, such as near schools and on the approach to sharp curves. Finally, "dynamic" systems derive real-time traffic information to warn the driver or limit vehicle speed at lower speed thresholds in poor weather, on slippery roads, around major incidents, and so on. Versions of these systems have been and are being tested in a number of countries (see Regan et al., 2003, for a review).

In summary, ISA is an example of a technology, likely to be deployed soon in production road vehicles, that is capable of duplicating the posted speed limit on a display within the vehicle cockpit and warning the driver if it is exceeded. It can go a step further and automatically limit vehicle speed based on real-time knowledge of transient changes in speed limits and prevailing traffic, road, and weather conditions.

In principle, the same combination of vehicle-based and cooperative ITS technologies brought together to build ISA systems can be used to display to the driver, within the vehicle, other messages currently presented on regulatory road signs. For example, messages displayed on signs regulating vehicle movement (e.g., "Stop,"

"Give Way"), vehicle direction (e.g., "No Left Turn"), and parking behaviors ("No Standing," "Loading Zone," "Clearway") could be repeated to the driver via an in-vehicle display. As with ISA, one can image the development of systems that intervene along a continuum, from those that merely duplicate road signage to those that, for example, physically intervene to slow the vehicle on the approach to a "Give Way" and/or "Stop" sign, prevent the car from making a left turn, or prevent the car from driving in a clearway zone. The same technologies can be used to display to drivers and pilots within a vehicle the messages and warnings on signs found in other transport domains.

Certain technical and logistic issues, however, will need to be addressed before information currently displayed on regulatory signs, including speed limit information conveyed by prototype ISA systems, can be reliably duplicated and presented to the driver within the vehicle cockpit. For example, all of the prototype ISA systems tested to date in large-scale field studies are of the "fixed" type (Carsten and Tate, 2000); that is, the onboard digital map is not capable of being automatically updated in real time about when and where a transient change to the posted speed limit (e.g., in the vicinity of the VMS or temporary road work) occurs. The road infrastructure and reference architecture to enable this to happen is yet to be developed in most jurisdictions around the world. Although it is possible to use transmitters located on or near existing regulatory signs to transmit information on them to a receiver in the vehicle, this is a costly and less practical option in the foreseeable future compared to using digital map technology.

Next, changes to speed and other regulatory signs are often made by local governments to meet their own specific requirements. When this occurs, updating the digital maps residing in vehicles is necessary. This may not be possible for weeks or months, depending on how long it takes local authorities to notify state or national road authorities of them and how long it takes the latter authorities to respond to these notifications.

Finally, a problem with most existing ISA systems that rely on GPS/digital map technologies is that sometimes the speed limit for side roads is incorrectly identified as the speed limit for the road on which the ISA-equipped vehicle is traveling. This is a function of the variation in GPS accuracy, which can occur hour by hour and can be avoided only by use of highly sophisticated and costly GPS equipment that uses differential correction. The costs for such systems at present would be prohibitive for use in the normal road transport industry. In Europe, several projects are underway, most of them funded by the European Commission (EC), aimed at increasing the accuracy of satellite-based vehicle positioning (the Galileo and EMILY projects), increasing the accuracy of digital maps (the NextMAP project), and developing standardized mechanisms for keeping digital maps up to date (ERTICO — ITS Europe, 2000). Similar large-scale initiatives are underway in North America and Japan.

14.5.2 Warning Information

Warning signs are used to alert drivers and other road users to potentially hazardous conditions on or adjacent to the road. They advise of conditions that require caution on the part of the driver and may call for a reduction in speed in the interest of safety of the driver and other road users (Standards Australia AS 1742.1, 1991). Like regulatory signs, warning signs typically fall into a number of subcategories:

- Alignment — e.g., "Road Curves Left".
- Intersections and junctions — e.g., "T-Junction".
- Advanced warning of traffic control devices — e.g., "Stop Sign Ahead".
- Road width and clearance — e.g., "Divided Road," "Left Lane Ends".
- Road obstacle — e.g., "Steep Descent," "Trucks," "Slippery," "Turning Traffic".
- Pedestrians/bicyclists — e.g., "Pedestrians," "School".
- Railway level crossing — e.g., "Rail Crossing".

As for regulatory signs, it is possible, in principle, to use the same combination of vehicle-based and cooperative ITS technologies employed by ISA systems to duplicate messages found on warning signs outside the vehicle on a display within the vehicle. The same technical and logistical challenges noted earlier, however, are relevant. Static warning signs are necessarily limited in the variety of warnings they can convey to the driver; they can alert drivers to permanent hazards, but they cannot warn the driver in real time of changing hazards.

In the following sections, some emerging technologies are reviewed that are capable of delivering to the driver within the vehicle a far greater range of warning information than can currently be presented by static warning signs or VMS outside the vehicle.

14.5.2.1 Traffic Information Systems

Variable Message Signs, as noted earlier, are capable of displaying to the driver real-time warning information that cannot be displayed using static warning signs. These signs are generally too expensive, however, to deploy in the vast numbers necessary to reflect all of the potentially hazardous conditions a driver might encounter. Cooperative ITS technologies are being developed for this purpose.

So-called traffic information systems (TISs), for example, convey information to the driver through various means. One is through the car radio via normal radio programs or dedicated channels. France, for example, has dedicated a radio frequency to motorway radio stations, allowing drivers to tune in anytime for updates on traffic jams, weather reports, alternative route suggestions, and so on (ERTICO — ITS Europe, 2000). In other jurisdictions in Europe, service providers now offer an automatic switch capability that moves the car radio receiver to a station broadcasting traffic information and then automatically returns it to the original station when the broadcast has finished. The radio offers many delivery options in addition to standard voice messages, including digital audio broadcasting (DAB), which, unlike FM radio broadcasts, can be used to transmit text and images at high speed, and the traffic message channel, which can deliver messages that can be filtered selectively and received in the driver's preferred language (ERTICO — ITS Europe, 2000).

Other TIS platforms for delivering traffic information to the driver have also been developed. In parts of Europe, for example, drivers can access certain Internet Web sites that provide information about roads on which travel is intended as well as route-planning options. Mobile telephones can be used to receive SMS messages that contain traffic information, and access to WAP (wireless application protocol) services allows the user to display traffic data on a digital mobile phone screen (ERTICO — ITS Europe, 2000). These kinds of technologies and services are adaptable for use in other transport domains.

14.5.2.2 Advanced Driver Assistance Systems

In addition to the technologies reviewed earlier, a number of other, predominantly vehicle-based, Intelligent Transport Systems capable of warning the driver in real time of a range of potential or actual traffic hazards have emerged. These are usually referred to as advanced driver assistance systems (ADAS) and are capable of alerting drivers within the vehicle to a much wider range of traffic hazards than can be reflected on static warning signs or the VMS. Many such systems exist, and it is beyond the scope of this chapter to review these in detail (see, however, Regan et al., 2001, and Rumar et al., 1999, for reviews). Vehicle-based ITS systems, for example, have been developed that are capable of warning drivers, via displays inside the vehicle, if:

- They are exceeding the speed limit or some other speed threshold (e.g., intelligent speed adaptation).
- They are about to collide with an obstacle in their forward path (forward collision warning).
- They are following a vehicle ahead too closely (following distance warning; adaptive cruise control).
- They are about to reverse into an obstacle behind them (reverse collision warning).
- They are about to collide with an object in their blind spot when changing lanes or merging (lane change collision warning).
- They are about to drive off the road (lane departure warning).
- Their cognitive, perceptual, and motor abilities have become impaired (e.g., drowsiness detection system; alcohol detection system).

A related category of vehicle-based ITS has been developed that is designed to discourage or prevent drivers from being exposed in the first place to traffic-related hazards of various kinds. Technologies in this category include:

- Systems that remind vehicle occupants to fasten their seatbelt and/or prevent the vehicle from being started if the driver and/or any other vehicle occupant is unrestrained (seat belt reminder; seat belt interlock).
- Systems that passively, or actively via on-board breathalyzer units, detect that a driver has been consuming alcohol and warn the driver not to drive and/or prevent the vehicle from being started if the driver's blood alcohol

concentration (BAC) is over some predefined threshold (alcohol sniffer; alcohol ignition interlock).
- Systems that detect an unauthorized or unlicensed driver and issue a warning to the driver and/or prevent the vehicle from being started (electronic license).

Cooperative ITS technologies enable presentation of an almost unlimited number of additional warnings to the driver within the vehicle or on a VMS outside the vehicle. The amount of information presented to the driver is constrained only by the existing limitations of the technologies used to collect traffic-related data and transmit them to the vehicle and, of course, the limited information-processing capacity of the driver. Cooperative ITS systems have been developed and deployed, mainly in Japan, that are additionally capable of warning drivers if:

- They are about to collide with a wide range of forward obstacles, including pedestrians crossing the street.
- They are about to collide with another vehicle when driving through, or turning at, an intersection.

The boundaries of cooperative ITS are ever expanding. As noted earlier, it is now possible to use everyday devices, such as mobile phones and pocket PCs, as components of cooperative applications. Again, the kinds of technologies and services reviewed here are adaptable for use in other transport domains.

14.5.3 Guide Information

Guide signs inform and advise road users about the route that they are following and give directions and distances to destinations on the route or along intersecting roads. They also supply information to enable road users to identify points of geographic or historic interest and give directions to motorists about the location of services, tourist facilities, and attractions (Standards Australia AS 17421.1, 1991). These signs include those relating to:

- Advanced direction — e.g., Roundabout Diagrams.
- Intersection direction — e.g., "Left Melbourne, Right Sydney".
- Reassurance direction — e.g., "Canberra 300/Sydney 1000".
- Street names — e.g., "Smith Street".
- Geographic features — e.g., "Swan River".
- Services — e.g., "Information 300 m on Left".
- Route markers — e.g., "Highway 31".
- Traffic instructions — e.g., "Slow Vehicle Lane Ahead".
- Kilometer posts — e.g., "W 30".
- Tourist information — e.g., "Scenic Lookout 300 m on Left".
- Freeway guidance — e.g., "Berrima Exit 2 km".
- Freeway service — e.g., "Information Bay 2 km".
- Freeway instructions — e.g., "Exit Speed 65 km/h".

Several technologies have been developed and deployed to provide drivers, via displays within the vehicle, with route guidance and navigation-related information that might otherwise be presented on road signs. Usually, this is achieved by combining traffic information services with an in-vehicle navigation system (IVNS). Route guidance can be on- or off-board and dynamic or static (ERTICO — ITS Europe, 2000). An off-board, static system, for example, can be used for pretrip planning. Here, the driver can use a personal computer (PC) to log onto a Web site to obtain directions for the quickest route to a chosen destination. Dynamic Internet-based systems take into account, in real time, prevailing traffic and weather conditions in calculating routes.

Early generation in-vehicle navigation systems are an example of an on-board, static system. These systems were one of the earliest ITS technologies to be deployed in production vehicles. The IVNS basically works as follows. After the driver programs a chosen destination (e.g., 10 Smith Street, Melbourne) into the system via the user interface (manually or using voice commands), the system calculates the shortest or fastest route to the destination and issues, via a visual display and/or a speech-based interface within the vehicle, turn-by-turn instructions on how to reach the destination (e.g., "In 200 m, Turn Right").

The basic operating components of these systems are a display interface (visual or speech-based), computer, digital road map database, and input sensors that supply information such as vehicle position (usually using the GPS) (ERTICO — ITS Europe, 2000). Dynamic IVNSs give the driver real-time access to traffic flow information on his chosen route to help him make informed decisions about alternative routes, departure times, or modes of transportation. A great deal of additional information can be derived and overlaid on IVNS displays. Modern digital map databases incorporated into these systems contain information about the topology and geometry of the road network, classes and types of roads, road restrictions, driving directions, street names, bridges and tunnels, administrative and state borders, and points of interest. The digital maps in these systems are generally more detailed than those employed for intelligent speed adaptation.

The latest model IVNS technologies allow drivers to take advantage of existing technologies such as GSM telephones, the pocket PC, and a remote service provider, without the need to purchase a built-in, onboard system. For example, the user can select a destination on the Internet using a Web-based interface. The Web-based system generates destination directions that can be downloaded onto a pocket PC.

Alternatively, the PC can be programmed directly. The computer, which doubles as a portable navigation display, uses the GSM telephone within the vehicle to send these destination data to a remote service provider, who computes the fastest or shortest route (whichever is preferred), taking into account the current traffic situation and bypassing traffic jams, road blocks, and detours. The route can be recomputed, if necessary, while the driver is en route to the destination. The great advantage of these systems is that drivers do not need to have a digital map of the road network onboard the vehicle, nor is it necessary to purchase map updates (ERTICO — ITS Europe, 2000).

As for regulatory and warning signs, it is possible to overlay the full range of messages currently found on road signs outside the vehicle on the IVNS display. Indeed, much of the information currently found on road signs is already duplicated on IVNS displays. In recent years, it has become obvious to manufacturers that digital maps can be used for many purposes other than in-vehicle route guidance and navigation. Highly accurate maps containing precise information about road geometry and related attributes in front of the vehicle could be used, for example, to augment advanced driver assistance systems by providing the driver with advanced warning of sharp curves ahead, forward obstacles, road narrowing, steep descents, and so on. If the digital map is accurate and up to date and the vehicle positioning system is sufficiently accurate, virtually any feature of the road environment may be pinpointed and used to trigger appropriate regulatory, warning, and/or guide messages to the driver.

In summary, as for regulatory and warnings signs, a trend is emerging toward the presentation to the driver within the vehicle of traffic information to assist him in planning and navigating his way to a chosen destination. Collectively, these emerging technologies are capable of presenting a greater variety of information than is currently displayed by traditional guide signs or the VMS. All of these technologies and services are adaptable for use by drivers and pilots in other transport domains.

14.6 CONCLUSIONS AND IMPLICATIONS

In summary, many ITS technologies exist and are being developed that make it possible to present regulatory, warning, and guidance information currently presented on static road signs outside the vehicle to the driver inside the vehicle. These systems are additionally capable of alerting the driver, in real time, about changing traffic conditions. Some of these ITS technologies are vehicle based, some are infrastructure based; and some are cooperative, involving communication between the vehicle and the road infrastructure and between the vehicle and other vehicles. Collectively, these systems are capable of providing the driver with a far greater variety of traffic information than can currently be displayed on static road signs. In many cases, it is possible to link these systems to vehicle control systems so that if a driver fails to react to a warning, the system automatically responds to the warning by slowing the vehicle or controlling it in the appropriate manner.

Of course, if the traffic information presented by these systems is to be detected, perceived, understood, and acted on in the intended manner by drivers and other road users, it is critical that it be designed and presented in a manner compatible with human information-processing capacities and limitations and be packaged in a way acceptable to them. If it is not, the information presented could potentially confuse, overload, and distract the driver.

In the next chapter, the human factors implications of the technological developments reviewed in this chapter are discussed.

REFERENCES

Carsten, O. and Tate, F. (2000). External vehicle speed control. Final Report: Integration. Leeds, U.K.: Institute for Transport Studies, University of Leeds.

ERTICO — ITS Europe/Navigation Technologies (2000). ITS — Part of everyone's daily life. Brussels, Belgium: Intelligent Transport Systems and Services — Europe (ERTICO)/Navigation Technologies.

Regan, M., Oxley, J., Godley, S., and Tingvall, C. (2001). Intelligent Transport Systems: safety and human factors issues. Royal Automobile Club of Victoria Literature Report No. 01/01. Melbourne, Australia: RACV Ltd.

Regan, M.A., Mitsopoulos, E., Triggs, T.J., Tomasevic, N., Young, K., Healy, D., Tierney, P., and Connelly, K. (2002). Evaluating in-vehicle Intelligent Transport Systems: a case study. *Proc. Road Saf., Res., Policing Educ. Conf.*, Adelaide, Australia, pp. 451–456.

Regan, M.A., Young, K., and Haworth, N. (2003). A review of literature and trials of intelligent speed adaption devices for light and heavy vehicles. Austroads Report No: AP-R237103. Sydney, Australia: Austroads.

Rumar, K., Kildebogaard, J., Lind, G., Mauro, V., Berry, J., Carsten, O., Heijer, T., Kulmala, R., Machata, K., and Zakor, I. (1999). Intelligent transportation systems and road safety. Brussels, Belgium: European Transportation Safety Council.

Standards Australia (1991). Australian Standard AS 1742.1-1991. Manual of uniform traffic control devices, Part 1: general introduction and index of signs. Sydney, Australia: Standards Australia.

Winnett, M.A. and Wheeler, A.H. (2002). Vehicle-activated signs — a large scale evaluation. Transport Research Laboratory (TRL) report 548. Berkshire, England: TRL Limited.

15 A Sign of the Future II: Human Factors

Michael A. Regan

CONTENTS

15.1 Introduction 225
15.2 Scenario 1 — Signs and VMS 226
  15.2.1 Potential Advantages 226
  15.2.2 Issues 226
15.3 Scenario 2 — Signs, VMS, and In-Vehicle Displays 227
  15.3.1 Potential Advantages 228
  15.3.2 Issues 229
15.4 Human–Machine Interface Design 230
  15.4.1 Design Guidelines and Standards 230
  15.4.2 Workload Managers 232
  15.4.3 Display Location 232
  15.4.4 Driver Acceptance 233
15.5 Procedural Guidance 234
15.6 Selection and Training 235
15.7 Summary 235
15.8 Will We Ever Do Away with Traffic Signs? 235
15.9 Conclusion and Implications 237
Acknowledgments 237
References 237

15.1 INTRODUCTION

The previous chapter served to illustrate alternative means by which traffic information can be conveyed to road users through the use of existing and emerging ITS (intelligent transport system) technologies, primarily from the perspective of the road vehicle driver. To summarize, these developments will make it possible in the near future to present the average driver with traffic information from several sources:

- A traditional, static traffic sign.
- A VMS displaying regulatory, warning, and guide information found on existing traffic signs.

0-415-31086-5/04/$0.00+$1.50

- A VMS device displaying traffic information that changes over time to reflect real-time traffic conditions.
- An in-vehicle device that displays traffic information found on existing static road signs and VMS — referred to as augmented signage (Campbell et al., 1998).
- An in-vehicle device displaying traffic information derived from other sources, such as onboard ITS technologies, that changes over time to reflect real-time traffic conditions.

At this early point in the evolution of ITS technologies, most vehicle drivers in countries that have deployed ITS technologies are exposed simultaneously to the first three sources (with only a small minority exposed additionally to some information from the last two sources). This will be referred to as scenario 1. As more ITS-equipped vehicles penetrate the market and the cooperative technologies to support them are further developed and deployed, however, a time will come during which most drivers in these countries will become exposed to all five sources of traffic information — scenario 2. Ultimately, if the static road sign becomes obsolete, drivers will be exposed to information only from sources 2, 3, 4, and 5. This might be referred to as scenario 3 but is considered unlikely for reasons discussed at the end of this chapter. In this chapter, the human factors implications of the technological developments that characterize scenarios 1 and 2 are discussed. The chapter concludes with a consideration of whether, in the future, roadside traffic signs will be needed.

15.2 SCENARIO 1 — SIGNS AND VMS

In this scenario, the primary sources of traffic information are the static road sign and the variable message sign (VMS) .

15.2.1 Potential Advantages

The main advantage of the VMS over traditional traffic signs is that it can present to the driver real-time traffic information that cannot be displayed on traditional, static traffic signs. It can be used to display regulatory, warning, and guidance information and, with its signal-like properties, is good at attracting driver attention. Other advantages of the VMS over the static sign are discussed in earlier chapters of this book. The main advantage of the VMS over systems that present traffic information within the vehicle is that the information on it is accessible to all drivers, regardless of whether they drive ITS-equipped vehicles. The positive safety benefits deriving from use of the VMS have been well documented (Rumar et al., 1999; Winnett and Wheeler, 2002).

15.2.2 Issues

The usual human factors considerations relating to the design of static road signs must be considered in designing VMS signs and the messages they convey to ensure that they can, usually in a single glance, be detected, read, understood, and acted on in the intended manner by a diverse range of road users. These were discussed in earlier chapters of this book. The VMS is, however, a much more flexible display device than a traditional traffic sign. Like a computer monitor, it is capable of displaying information in a variety of formats, in a range of colors, and at a range of different intensities. It is important, therefore, that relevant human factors guidelines and standards developed to inform the design of the human–computer interface for advanced traveler information systems are consulted by traffic engineers in designing information displayed by the VMS (see Campbell et al., 1998).

A dilemma inevitably faced by traffic engineers is whether to (1) duplicate the same information on the VMS using the same coding schemes (e.g., shape, color, verbal/symbolic) and graphics (letter shapes, symbol design) currently displayed on static signs or (2) develop new display designs more effective in facilitating information transfer and thus take advantage of the flexibility offered by the VMS interface. This dilemma was faced decades ago by the designers of aircraft cockpits when old analog cockpit displays were replaced by electronic cathode ray tube (CRT) displays. The new-look "glass cockpits" could be designed to mimic electronically the old analog dials and gauges in earlier cockpits or to display information to pilots in a more effective, integrated manner, utilizing new colors, graphics, and coding schemes. The first-generation glass cockpits merely duplicated electronically the old analog displays. Later-generation cockpits, however, incorporated original and modified displays.

Deciding which approach to take is necessarily a tradeoff between maximizing information transfer, on the one hand, and extinguishing some long-established stimulus–response patterns derived from years of exposure to information on traditional road signs, on the other. User acceptability is another critical issue to consider; even now, many pilots in commercial aviation are reluctant to upgrade to glass cockpits because they do not perceive the new cockpits to be as challenging and user friendly as older-generation cockpits.

In the present context, it is most critical to ensure that, whatever approach is taken, the design of information between VMS and VMS, and between VMS and existing static traffic signs, is compatible across the road network. This is essential to avoid confusion in the minds of road users and ensure that all signs are responded to in the manner intended. The primary compatibility considerations here relate to the coding and color schemes, fonts, and layouts associated with the various categories of traffic information displayed.

15.3 SCENARIO 2 — SIGNS, VMS, AND IN-VEHICLE DISPLAYS

The issues become more complex in the second scenario. Here, the driver of an ITS-equipped vehicle will be exposed to traffic information deriving from several sources: existing traffic signs, VMS, and displays within the vehicle. In this scenario, information from inside and outside the vehicle potentially can impinge simultaneously on the driver.

15.3.1 Potential Advantages

In scenario 2, the information presented from within the vehicle can take the form of augmented signage (i.e., information found on static signs or the VMS that is duplicated, in the same or a different format, inside the vehicle) or real-time traffic information derived from other sources, such as vehicle-based ITS. Several potential advantages are associated with the provision to drivers within the vehicle of augmented signage:

- Information can be presented in multiple, redundant, sensory modalities (visual, auditory, tactile, and olfactory) to maximize the likelihood that it is detected, comprehended, and acted on in the intended manner.
- Presentation of information via sensory modalities other than vision can free up an already heavily burdened visual system; it has been estimated, although with some dispute, that over 90% of traffic information processed by the driver is visual (Hills, 1980).
- More detailed instructions can be conveyed to the driver than can be extracted from a roadway sign in a single glance.
- Messages can be captured, held, and displayed for as long as is needed for the driver to comprehend and act on them in the manner intended.
- Messages can be presented earlier in time to the driver than they can on traffic signs and VMS so that he has more time to act on it, although this might encourage premature braking or other premature responses possibly counterproductive from a safety perspective (Campbell et al., 1998).
- It is possible for the driver to repeat on command messages that have not been initially attended to, comprehended, or obscured by road furniture or objects.
- Information can be read and comprehended more reliably under adverse weather conditions, at different times of day, when driver visual acuity is limited, when driving into the sun, and so on (Campbell et al., 1998).

Another advantage associated with the provision of augmented signage within the vehicle is that it can be optimized for use by drivers with special needs, such as young novice drivers, older drivers, and the functionally impaired, resulting in better information transfer than would be possible with external signs alone. This might be achieved in various ways. For example:

- Information could be made available in alternative, driver-selectable, sensory modalities; this would allow visually and hearing-impaired drivers to adapt presentation of the information to their perceptual abilities (Campbell et al., 1998).
- Information could be broken down into smaller information units (Campbell et al., 1998).
- Drivers could be allowed to filter information presented to them inside the vehicle to receive only that required to complete their journey safely;

as it is, drivers tend to look only at signs relevant to their trip, no matter how conspicuous other signs are (Theeuwes, 2002).

- The system could provide means for the driver to enlarge information displayed visually in order to adapt the visual output to his needs (Campbell et. al., 1998).
- Information could be given in a choice of verbal or pictorial formats presented in the driver's own language and, in the case of pictorial signs, in the same format used in the driver's native country.
- Information could be designed to provide clear guidance to the user as to what he should do in order to respond to the information safely (for example, guidance on how to navigate through the roadwork).

The preceding advantages apply generally to any traffic-related information that can be displayed to the driver internally and externally to the vehicle, whether it is static or dynamic. As Campbell et al. (1998) note, however, the critical task for designers is to provide augmented signage when it is needed and avoid distracting the driver when it is not. Just as critical is deciding what information drivers should be allowed to filter out selectively in order to protect them from information overload. It would be unwise, for example, to allow them to filter regulatory information selectively.

It is possible to confer some of these advantages on the driver through changes to the design of existing external signage. As discussed earlier in this book by Kline and Dewar, for example, much interest at present concerns the way in which the traffic environment can be optimized for older drivers, given that, in most developed countries, the number of drivers over the age of 65 is expected to double in the next 40 years and drivers in this group are overinvolved per kilometer traveled in fatal and serious injury crashes. In North America, this has led to the U.S. Federal Highway Administration's development of an older driver highway design handbook (Staplin et al., 2001).

This manual identifies four types of roadway sites as most problematic for older drivers: at-grade intersections, grade separation intersections, roadway curvature, and passing zones and highway construction zones. It makes recommendations for addressing these problems. For static signs, these include recommendations to make messages on them more visible and conspicuous to elderly drivers, to provide multiple advanced warning of intersections, and to provide positive guidance in traffic situations problematic for the older driver. For the VMS, it is recommended that no more than two phases be used, with no more than three units of information presented for each phase. Similar recommendations have been derived from research undertaken in Australia (Fildes et al., 2000). Although these changes to external signage are designed to make life easier for the older driver, they will, of course, benefit drivers in general.

15.3.2 Issues

Presenting the driver simultaneously with multiple messages and warnings — from static signs and VMS outside the vehicle as well as from displays inside the vehicle — has the potential to overload, distract, and confuse the driver. Overload can occur if the combined information load from internal and external sources exceeds the limited processing capacity of the driver. Incompatibilities in the design and timing of information presented inside and outside the vehicle can lead to driver confusion, distraction, and overload. Poorly designed and/or located displays inside the vehicle have the potential, likewise, to degrade the performance of the driver.

Traditionally, three human factors approaches can be taken in scenario 2 to minimize these potentially negative consequences: (1) good ergonomic design of the human-machine interface, (2) good procedural guidance in how to interact optimally with the interface, and (3) driver selection and training. These are considered in Section 15.4.

15.4 HUMAN–MACHINE INTERFACE DESIGN

It is critical that the HMI is designed in accordance with good ergonomic practice. Some of the key issues to be considered are discussed in this section.

15.4.1 Design Guidelines and Standards

Handbooks already exist that summarize human engineering data, guidelines, and principles to assist vehicle cockpit designers and traffic engineers in designing the human–machine interface to support the driver with information deriving from emerging ITS technologies from within the vehicle and on VMS (Campbell et al., 1998). In Europe, North America, and Japan, standards are also being developed that contain performance-based goals to be reached by the HMI so that onboard systems do not, among other things, distract or visually entertain the driver while he is driving. The European Commission, for example, has issued a European Statement of Principles for the design of the HMI for in-vehicle information and communication systems (European Commission, 2001). The Alliance of Automobile Manufacturers (AAM) in the U.S. has issued a similar document (AAM, 2002).

These standards leave it open to system designers to use good ergonomic practice in designing the HMI, without creating unnecessary obstacles or constraints to innovative development of products and systems. The International Organization for Standardization (ISO), through its various subcommittees (in particular Subcommittee 13 of Technical Committee 22), is also developing draft standards for the design of the HMI for transport information and control systems (TICS) that, if adhered to, will result in some degree of standardization of HMI design across vehicles. Checklists have also been developed for assessing the ergonomic quality of the HMI for transport information and control systems (Stevens et al., 1999).

The various guidelines and standards under development are concerned with the design, presentation, and prioritization of traffic-related information presented to drivers within the vehicle. If, during scenario 2, most drivers are presented with traffic information from inside and outside the vehicle, it will be necessary to develop guidelines for the design of what might be called the human–machine–environment interface (HMEI). These would need to take into account all sources of traffic-related information that the driver will be expected to detect, comprehend, and act on from within the vehicle and outside it.

Such guidelines would need to address several issues: the overall information load deriving simultaneously from within and outside the vehicle, compatibility of messages presented to the driver inside and outside the vehicle, relative timing of presentation of such messages, and prioritization of messages deriving simultaneously from within and outside the vehicle. The most critical issue to resolve in the first instance is who, ultimately, is responsible for optimizing the design and presentation of traffic information presented to the driver from inside and outside the vehicle. It is neither the road designer alone nor the vehicle designer alone. The human factors specialist, as the interface between the two, is in a special position to initiate dialogue between these two professional groups who, historically, have had little to do with each other.

The development of more encompassing guidelines and standards of this kind will, however, be a challenging task because many empirical issues need to be resolved. Can it be assumed, for example, that if augmented signage is provided within the vehicle, the driver will no longer pay attention to road signs conveying the same information outside the vehicle? After all, drivers have become conditioned through years of driving to look at and attend to information on traffic signs that they regard as relevant automatically. In scenario 2, the net effect of providing augmented signage within the vehicle may be an increase rather than a decrease in driver workload and distraction. Alternatively, it is possible that providing traffic information within the vehicle may lead to over-reliance on the information presented internally. Thus, drivers may become less inclined to scan the external road environment for traffic information on signs in the sometimes false belief that their internal displays have given them all the necessary information to drive legally and safely.

Timing of presentation of augmented signage is a critical issue. A potential advantage of augmented signage within the vehicle is that it can be presented earlier to the driver. The few studies conducted on this issue to date suggest that earlier presentation of information inside the vehicle, relative to the same information presented externally, results in faster driving responses; this has usually been interpreted as a benefit as Campbell et al. (1998) point out. However, this may not always be the case; in many driving situations, braking or reacting in some other way prematurely might actually cause a crash. The message must be presented early enough for the driver to respond but not so early that the driver brakes or reacts prematurely or inappropriately and causes a crash (Campbell et al., 1998). At a macro level, the issue is more complex. Drivers are used to the way other drivers drive and react to traffic information and they adjust their own driving performances accordingly. Different drivers provided with the same traffic information but at different times — some externally and some internally — can potentially upset the equilibrium of the traffic system and, in doing so, compromise safety.

Finally, "augmented" signage need not simply involve the replication inside the vehicle of the same information found on existing traffic signs and on VMS. Augmented signage may take the form of redundant information presented in the vehicle, to make it more likely that external information is detected, comprehended, and acted on in the intended manner by the driver. Augmented signage may also be used, in a cooperative way, to supplement information found on existing traffic signs or VMS. For example, limited information on an external road sign might be used to prepare the driver for a more detailed message presented internally if this is found to be more effective in facilitating information transfer than only internal presentation of the message. Clearly, it is necessary to develop a taxonomy of the various options for presenting a combined suite of information to the driver from inside and outside the vehicle.

These and related issues must be resolved before standards and guidelines can be developed to optimize the design of traffic information presented to the driver from within and outside the vehicle.

Of course, if vehicles in the future are allowed to be retrofitted with ITS technologies as they come onto the market, performance-based standards will need to be developed to ensure that this is done so that it does not compromise driver safety. A related concern is the emerging trend toward the provision of traffic information services to drivers via portable technologies such as pocket PCs and mobile telephones. As noted in the previous chapter, it is now possible in some parts of the world, for example, to use a pocket PC as an in-vehicle navigation system. This creates a further challenge from a human factors perspective. Now, vehicle cockpit designers and traffic engineers need to communicate with each other to ensure that the human–machine–environment interface for the presentation of traffic information to drivers is ergonomically designed. Service providers must also be involved in these discussions to ensure that the information they provide through portable devices is compatible with other traffic-related information impinging on the driver from inside and outside the vehicle.

15.4.2 Workload Managers

A recent development in human–machine interface (HMI) design is the emergence of the so-called workload manager. This is a vehicle subsystem that utilizes data from onboard sensors and ITS technologies to monitor the vehicle and the environment around it and makes decisions about when information, including traffic-related information, should be presented to drivers. The general idea is to present information only when necessary, without overloading or distracting them from the primary driving task. This follows from initial concept development in the aviation domain of so-called pilot associates. The National Highway Traffic Safety Administration (NHTSA) in North America has recently funded a multiyear project known as SAVE-IT (safety vehicle using adaptive interface technology) to develop a prototype workload manager (Foley, 2003). As yet, no details about the research project are publicly available.

From the preceding discussion, it is clear that, to be effective, it will be necessary for these workload managers to take into account all sources of traffic information impinging on the driver: that presented within the vehicle and that presented externally.

15.4.3 Display Location

If well designed, external signs and VMS do not usually draw the driver's eyes away from critical events in the forward line of sight. Displaying traffic information visually to the driver inside the vehicle, however, is problematic in most current vehicles. In most, the driver must look inside the vehicle and thus away from the road in order to read and comprehend information displayed visually — even if the displays are mounted high in the vehicle cockpit. It will be necessary to present any greater volumes of information to be displayed visually to the driver in a manner that does not further increase eyes-off-the-road time.

The motoring world has shown much enthusiasm for introduction of the automotive head-up display (HUD). The HUD, like that in modern military and some commercial aircraft, is an electro-optical device that superimposes static and dynamic symbols and graphics on the vehicle windshield within the driver's forward field of view. The image is optimally collimated to infinity to reduce the need for the driver to accommodate the eyes again when shifting his point of focus from the displayed image to images in the outside world. The driver is thus able to sample vehicle and driving information without the accommodative shift required by conventional head-down instrument clusters (Campbell et al., 1998). The HUD can also be combined with an infrared capability to enable the driver to "see in the dark" through the windshield. Although first-generation HUD systems already exist in some luxury vehicles, their ability to improve driving performance and safety has been much debated (see, for example, Regan et al., 2001). A primary concern is that images on the display may obscure or distort the shape of objects in front of the vehicle, such as road signs and pedestrians.

The potential problems associated with HUD technologies will need to be resolved before it is safe to increase the amount of traffic information displayed visually inside the vehicle. In the meantime, it would seem fruitful to devote more human factors research effort to better understand and exploit advantages associated with the presentation of information to drivers through sensory modalities other than the visual modality, given that emerging technologies now make this possible. In choosing to exploit these advantages, however, the designer must be careful not to create new problems by overloading other sensory modalities or unnecessarily extinguishing long established stimulus–response patterns that facilitate drivers' visual comprehension of existing traffic information.

15.4.4 Driver Acceptance

No matter how well designed the traffic information presented to drivers, it will not be acted on in the intended manner unless they deem it acceptable. Auditory messages presented within the vehicle that, for example, interfere with music deriving from entertainment systems will not be acceptable to drivers unless they are carefully human engineered. To be acceptable, the traffic information presented to drivers must be user friendly, useful, effective, socially acceptable, and affordable (Regan et al., 2002; Young et al., 2003). Some have further argued that, to be acceptable, the HMI must also be "pleasurable" (Jordan, 1999). A gentle "chime," for example, is a more pleasurable sound than a "screeching" sound. If the traffic information presented to the driver from within the vehicle is unacceptable, he will ignore it, filter it out, or attempt to disable the equipment through which it is presented. If the technologies reviewed in the previous chapter of this book are to be successful in making drivers respond to traffic information in the intended manner, it is critical that acceptance testing involving end users is undertaken at all stages of system development.

15.5 PROCEDURAL GUIDANCE

Providing the vehicle driver with procedural guidance in how to interact optimally with the HMI is an issue that has received little empirical attention. Procedural guidance can be provided in many forms, such as in the form of a user manual. Allowing drivers to personalize the information they receive within the vehicle is a case in point. As noted previously, it will be possible in the future to design the HMI so that all information on external signs and VMS is displayed within the vehicle; drivers will be able to personalize it, as well as information derived from other sources, to suit their individual needs. This may include the facility to filter out traffic information that they regard as irrelevant to their journey. This raises some interesting issues.

A growing trend by vehicle manufacturers is to design the HMI in a manner that gives the driver considerable freedom to display information and operate internal devices in alternative ways to achieve the same functional outcome. Most modern cars, for example, offer several ways of tuning a radio to a particular frequency and selecting it as a station preset. The particular way chosen by a driver may not necessarily be the least cognitively demanding and distracting one while the vehicle is in motion (Young, Regan, and Hammer, 2003). Similarly, vehicles now exist that allow the driver to reconfigure, to suit his own preferences, the format of information on electronic display clusters within the cockpit. Some vehicles, for example, allow the driver to choose alternative formats for the speedometer (e.g., analog vs. digital) and other instruments.

In most cases, drivers do not read operating manuals to identify the most ergonomic means of interacting with in-vehicle technologies; even if they do, few, if any, manufacturers identify the option least likely to interfere visually, manually, and/or cognitively with the primary driving task. Giving the driver the option in the future of personalizing the presentation of traffic information may lead to negative safety outcomes for the individual if he does so in a less than optimal manner. At a macro level, this personalization may result in increased heterogeneity of vehicle behavior, which could adversely affect the overall safety and mobility of the road network.

Legal issues must also be considered in this context. As noted earlier, it might be undesirable to allow drivers to filter out certain traffic information, such as that presented on regulatory signs. Allowing them to do so might result in them receiving traffic violations or even crashing (Campbell et al., 1998).

In summary, very little research has been done to define in-vehicle information presentation design parameters that could, and should, be adjustable by drivers. When personalization of information presented within the vehicle is allowed, options for doing so should be restricted to those that will not compromise driver safety. Procedural guidance, in whatever form it is delivered, must be designed so that the driver actually refers to it and it supports the driver in personalizing information in the safest manner possible.

15.6 SELECTION AND TRAINING

Traditionally, the tools of selection and training have been used, in part, by the human factors engineer to moderate the degree of mental workload experienced by an individual. People can be selected to operate complex vehicles on the basis of their performance on tests predictive of their ability to operate those vehicles. Alternatively, or in addition, people can be given formal training to develop skills that enable them, through practice, to perform complex tasks more efficiently and more safely than they otherwise could do without training.

As road vehicle interfaces become more complex, it is likely that drivers will require at least some degree of formal training and education to operate them. The purpose of this will be to:

- Develop the perceptual, cognitive, and manual control skills needed to operate such vehicles.
- Explain limits to the capabilities and limitations of the onboard systems.
- Demonstrate to users the intended use of the systems.
- Demonstrate to users optimal modes of interaction that minimize distraction and workload.

While nonequipped vehicles remain in the vehicle fleet, it will also be necessary for drivers to undertake traditional training so that, in the event that they need to drive a nonequipped vehicle, they can do so. Through the licensing system, untrained drivers, or drivers assessed as incapable of operating vehicles equipped with certain ITS technologies, might be restricted from operating certain vehicles. At the present time, such approaches may not be palatable to a society in which people are generally allowed to drive any make or model of car, without the need for any specific formal training.

15.7 SUMMARY

The ability to display traffic information to drivers from inside and outside the vehicle raises some important human factors issues. If well designed, the human–machine–environment interface through which the information is presented has the potential to increase the driver's situational awareness of the traffic system, thus increasing safety and transport efficiency. If poorly designed, however, it has the potential to overload, distract, and confuse the driver. The human factors issues considered here are generally relevant to the design and deployment of Intelligent Transport Systems used to convey traffic information in a similar manner to operators in other transport domains.

15.8 WILL WE EVER DO AWAY WITH TRAFFIC SIGNS?

Earlier in this chapter a final scenario, scenario 3, was envisaged in which the static road sign might become obsolete. It is reasonable to ask whether, given the developments in traffic technology reviewed in the previous chapter, traditional traffic signs will be necessary in the future.

Certainly, as noted earlier, it is already possible through automation to make vehicles behave, without human intervention, in a manner that automatically conforms to messages displayed on some regulatory signs. Limiting ISA systems is one such example. At least in theory, these systems render the traditional speed sign obsolete. However, at the present time, these systems are not reliable in all circumstances and many drivers find the concept of vehicle control being taken away from them, partially or fully, unacceptable (Regan et al., 2002). Drivers who accept such technologies must have a high level of trust in the system. For drivers to be able to assess the reliability of such systems, it is probable that they will, for quite some time, need road signs as a feedback mechanism to enable them to judge the reliability of a particular system.

Other reasons exist for believing that the traditional traffic sign will be around for some time to come. Certainly, although road network vehicles not equipped with ITS technologies and capable of presenting traffic information to the driver inside the vehicle remain on the road, traffic signs and the VMS will continue to be necessary. Similarly, road users other than vehicle drivers, such as bicyclists and pedestrians, rely on traditional traffic signs to navigate their way legally and safely through the road network. Interestingly, pedestrians in Japan and North America now have access to real-time navigation assistance by accessing Internet services via a mobile telephone. The telephone screen can be used to display route information in the form of simplified maps and/or text instructions. As in vehicles, it is likely in the near future that the capabilities of these systems can be extended to provide pedestrians and other road users, via portable devices and alternative sensory modalities, with a much wider range of traffic-related information than is currently found on road signs. It will be some time, however, before all road users have access to such devices.

Even if road users are able to access traffic information remotely, within the car, or using mobile display devices, there is no guarantee that these systems will operate reliably at all times. Also, there is no guarantee that remote devices will be operated in the manner intended by the designer and that users will remember to maintain them in order to receive reliable information. From a duty-of-care perspective, the traditional road sign has been proven a reliable platform for assisting the road user to navigate through the road system legally and safely.

Taking away signs altogether in a future motoring world dominated by automation raises an interesting legal issue: in the event that a vehicle, due to some malfunction of an ITS system, failed to respond at a given location in a manner consistent with a message on a regulatory sign previously displayed at that location (e.g., STOP), who is legally liable in the event of a crash — the driver, vehicle manufacturer/supplier, service provider, or road authority? In the past, the onus of liability has been on the driver when failing to stop at a STOP sign. In a world without STOP signs, the situation would not be so clear. This alone may provide a sufficient reason for society to retain traditional traffic signs.

In summary, it is unlikely that the traditional traffic sign and the VMS will become obsolete in the near future, if ever. This will happen only if the information they convey can be presented more reliably, less expensively, and more safely within the vehicle than outside it. Many technical, ergonomic, legal, and social issues will need to be resolved before the demise of the humble road traffic sign.

15.9 CONCLUSION AND IMPLICATIONS

The concept of what is meant by a traffic sign is changing, driven by advances in traffic technology, considered here and in the previous chapter under the rubric of Intelligent Transport Systems. Although the focus in this and the previous chapter has been on traffic signs in the road transport domain, the issues raised are relevant to other transport domains.

The traffic sign of the future will not necessarily be a static or dynamic road sign attached to a vertical pole outside the vehicle. It may be located inside the vehicle and may take the form of a text or graphic image on a flat-panel display or HUD, a computer-generated word, or a vibration felt through the vehicle accelerator pedal. Alternatively, it may take the form of a remote, portable display implemented on a mobile telephone or other communication device. For the driver, the distinction between traffic information presented outside the vehicle and that presented inside it will become increasingly blurred.

These changes will make it possible to present information flexibly in a manner that facilitates information transfer, reduces road infrastructure costs, and supports the safety and mobility needs of particular driving subgroups such as the young and elderly. Unless the various methods for designing and presenting traffic information inside and outside the vehicle are carefully researched and implemented, however, the potential benefits to be derived from these emerging ITS technologies might be outweighed by their potential to overload, distract, and confuse the driver. Determining who is ultimately responsible for managing the evolution of the sign of the future is probably the most significant issue to be resolved in the short term.

ACKNOWLEDGMENTS

The author would like to thank Professor Tom Triggs, from the Monash University Accident Research Centre, for reading and commenting on an earlier draft of this chapter and the previous chapter.

REFERENCES

AAM (April, 2002). Statement of principles, criteria and verification procedures on driver interactions with advanced in-vehicle information and communication systems (Version 2). U.S.: Driver Focus-Telematics Working Group, Alliance of Automobile Manufacturers.

Campbell, J.L., Carney, C., and Kantowitz, B.H. (1998). Human factors design guidelines for advanced traveller information systems (ATIS) and commercial vehicle operations (CVO). Report No. FHWA-RD-98-057. McLean, VA: U.S. Department of Transportation, Federal Highway Administration.

European Commission (July, 2001). Safe and efficient in-vehicle information and communication systems: a European statement of principles on human machine interface. Brussels, Belgium: European Commission.

Fildes, B., Corben, B., Morris, A., Oxley, J., Pronk, N., Brown, L., and Fitzharris, M. (2000). Road safety environment and design for older drivers. Austroads Report No. AP-R169/00. Sydney, Australia: Austroads.

Foley, J. (2003). Personal communication.

Hills, B.L. (1980). Vision, visibility and perception in driving. *Perception*, 9, 183–216.

Jordan, P.W. (1999). Pleasure with products: human factors for body, mind and soul. In W. Green and P.W. Jordan (Eds.), *Human Factors in Product Design*. Philadelphia, PA: Taylor & Francis.

Regan, M., Oxley, J., Godley, S., and Tingvall, C. (2001). Intelligent Transport Systems: safety and human factors issues. Royal Automobile Club of Victoria Literature Report No. 01/01. Melbourne, Australia: RACV Ltd.

Regan, M.A., Mitsopoulos, E., Haworth, N., and Young, K. (2002). Acceptability of vehicle Intelligent Transport Systems to Victorian car drivers. Royal Automobile Club of Victoria Public Policy Report 02/02. Melbourne, Australia: RACV.

Rumar, K., Fleury, D., Kildebogaard, J., Lind, G., Mauro, V., Carsten, O., Heijer, T., Kulmala, R., Macheta, K., and Zackor, I.H. (1999). Intelligent transportation systems and road safety. Brussels, Belgium: European Transportation Safety Council.

Staplin, L., Lococo, K., Byington, S., and Harkey, D. (2001). Highway design handbook for older drivers and pedestrians. Federal Highway Administration Report No. FHWA-RD-01-103. McLean, VA: FHA.

Stevens, A., Board, A., Allen, P., and Quimby, A. (1999). A safety checklist for the assessment of in-vehicle information systems: a user manual (PA3536/00). Crowthorne, U.K.: Transport Research Laboratory.

Theeuwes, J. (2002). Sampling information from the road environment. In R. Fuller and J.A. Santos (Eds.), *Human Factors for Highway Engineers*. Amsterdam, Netherlands: Pergamon Press.

Winnett, M.A. and Wheeler, A.H. (2002). Vehicle-activated signs — a large scale evaluation. Transport Research Laboratory (TRL) report 548. Berkshire, England: TRL Limited.

Young, K., Regan, M.A., and Hammer, M. (2003). Driver Distraction: A review of the literature. Monash University Accident Research Centre Report No. 206. Melbourne, Australia: MUARC.

Young, K., Regan, M.A., Mitsopoulos, E., and Haworth, N. (2003). Acceptability of in-vehicle Intelligent Transport Systems to young novice drivers in New South Wales. Monash University Accident Research Centre Report No. 199. Melbourne, Australia: MUARC.

16 Author Reflections on the Human Factors of Transport Signs

Tim Horberry, Cándida Castro, and Patricie Mertova

CONTENTS

Cándida Castro (Book Co-Editor and Co-Author of Chapter 1, Chapter 4, and Chapter 16) 240
Tim Horberry (Book Co-Editor and Co-Author of Chapter 1, Chapter 4, and Chapter 16) 240
Maxwell G. Lay (Author of Chapter 2 and Chapter 3) 241
Francisco Tornay (Co-Author of Chapter 4) 241
Terry Lansdown (Author of Chapter 5) 242
Alexander Borodin (Author of Chapter 6) 242
Kirstie Carrick, Peter Pfister, Robert Potter, and Roy Ng (Co-Authors of Chapter 7) 242
Donald Kline and Robert Dewar (Co-Authors of Chapter 8) 243
Ray Fuller (Author of Chapter 9) 243
Hashim Al-Madani (Author of Chapter 10 and Chapter 11) 244
Luis Montoro, Antonio Lucas, and María T. Blanch (Co-Authors of Chapter 12 and Chapter 13) 244
Michael A. Regan (Author of Chapter 14 and Chapter 15) 245

In a contributor book of this kind, it is clear that multiple perspectives on the topic will emerge. We make no apologies for this and instead argue that it is a strength, in part due to the richness of work in this area. This final chapter is therefore intended to be "democratic" by allowing all of the chapter authors to write a paragraph or two about how they approached their topics.

The authors' reflections are presented in the order of their contributions to the book. Only minimal editorial changes have been made to their words.

0-415-31086-5/04/$0.00+$1.50

Cándida Castro (Book Co-Editor and Co-Author of Chapter 1, Chapter 4, and Chapter 16)

Much of the previously published research on traffic signs has focused almost exclusively on the issues of signing conspicuity and comprehension. As such, it neglected other valuable issues like attention, motivation, new technologies, international standardization, history, and signing materials. This book fills that gap by offering multiple perspectives on this topic. From my point of view, the main outcome of this book consists of gathering knowledge of traffic signs and information acquisition from the standpoint of human factors psychology, to help provide deeper insights into the way traffic users behave.

To take just one example, signs from different semantic categories are often employed to express the same message, for example, to enable the car driver to know which way he is allowed to go along a roadway (an obligation or a prohibition message on the sign). However, such different representations may be a key factor for subsequent significant differences in drivers' behavior.

As such, I argue that human factors psychology could be better applied to the design, manufacture, location, and maintenance of signals. Application of the content of this book could help avoid human-related transport problems, thus making signals more easily usable and intelligible. Driving a car, train, or forklift truck is inherently more hazardous than other everyday activities, such as using a computer. Therefore, the application of human factors knowledge to transport signing could not only help improve these devices but also help prevent accidents.

Hopefully, this book will make a useful contribution to safety in all transport domains. Without any doubt I enjoyed working on it and, although I understand its limitations, I believe that it contains a large amount of useful information to enrich knowledge in this field.

I want to sincerely thank all of the authors who have been involved in this project and have worked so hard to help this book come out. Thank you very much, Max, Patricie, Paco, Terry, Alex, Peter, Kirstie, Don, Bob, Ray, Hashim, Antonio, Luis, Maite, and Michael. I want also thank to Tim Horberry, my faithful co-editor, for working with me throughout the long process of putting this book together.

Tim Horberry (Book Co-Editor and Co-Author of Chapter 1, Chapter 4, and Chapter 16)

As the other editor and co-creator of this book, I find it difficult to separate my reflections of the chapters to which I contributed from my general reflections on the overall book. What do I think are important issues in the human factors of transport signs? Take a look at the table of contents in this book.

When Cándida Castro and I first conceived the idea of this book, we were each undertaking a Ph.D. in this general area; despite this, surprisingly little of our respective theses is actually contained in the information we present here. Hopefully this shows that the contents of this book are more than just academic information. Traffic signs should be a very fertile area for human factors professionals to work in; many well-established principles of engineering psychology, cognitive science, social psychology, visual design, and road user behavior are applicable.

However, as most of us have experienced when driving a car around a busy city center, the combination of factors such as vehicular control, interacting with other vehicles and, possibly, other people inside the vehicle, route planning, and filtering out irrelevant visual information is exceedingly complex. Of course, no single study can address all of them; even so, traffic sign research as a whole must take all of these factors into account for the overall findings to be realistic and usable by engineers and other transport professionals. Of course, this is no easy matter; we hope the information in this book will help integrate much of this research information.

Furthermore, the advent of Intelligent Transport Systems and other advanced computing and communication technologies in all transport domains will surely blur the line between information in the environment and information in the vehicle. As such, the nature and demands of the driving task are bound to change. However, safety, efficiency, human–machine integration, and pleasurability for the operator remain vital issues. Such things should keep human factors researchers busy for years to come.

Maxwell G. Lay (Author of Chapter 2 and Chapter 3)

In the first of my two chapters, I gave a short history of traffic signs, especially examining how road signs were influenced by rail and maritime factors. The second chapter, "Design of Traffic Signs," considers the purpose of traffic control devices, definition of signs, and material for signs. After this, traffic sign theory and traffic sign requirements are considered at some length. Readers who would like more information on these topics should consult my 1998 book, *Handbook of Road Technology.*

Francisco Tornay (Co-Author of Chapter 4)

I consider the present work to be a *springboard book.* It makes information readily available, without oversimplifications, but in a coherent way — as coherent as current knowledge allows. After reading the relevant parts, one can make calculated guesses about new improvements. If necessary, the book also makes it easier to go directly and with clearer ideas to the original relevant sources. The data presented directly suggest new lines of thought. One example: reading Chapter 4, I thought of the idea of using variable signs for introducing new signs for the same warning, which seems to increase drivers' compliance. Combining the new and old signs for some time may ensure that drivers understand the information. Better ideas would undoubtedly come from trained professionals. The next step is new research for assessing the effectiveness of such new ideas. This way, the gap between sound knowledge and new insights would be as small and controlled as possible. Only one final recommendation: there should be periodic updates of this book.

Terry Lansdown (Author of Chapter 5)

In drafting my chapter, the rationale adopted was to step back from the "nuts and bolts" of the size, color, and location considerations in the design of signage. The approach adopted was one in which a fundamental "rethink" is advocated in the assessment of the fitness for purpose of our signage systems. It is suggested that this approach should re-evaluate the need and provision of signage from the perspective of the driver's attentional capabilities. It is hoped that, by explicit consideration of the relationship between the individual's information-processing capabilities and the task and environmental demands, signage may be designed to better meet the user's requirements.

Alexander Borodin (Author of Chapter 6)

In writing my chapter, I have attempted to give a very broad description of railway signage and its use. In particular, my concern in writing it was that the reader gain a feel for the subject rather than an in-depth understanding of it. The reason for this is that the topic is quite vast, ever changing, and quite dependent on the particular railway and country one is examining. Thus, really, it is impossible to understand the topic fully.

The other area that I tried to give the reader an appreciation for was the fact that signage does not solve very many problems because people still ignore signs and signals — not because they hate signage designers or disagree with the principle behind the sign being there, but because the human condition is not one of slavish devotion to rules and regulations; it is one of conscious and subconscious thought and action. Human beings are not computers but highly complex, sentient, and intelligent beings. Therefore, signage systems need to develop to the stage that they serve humans and not the other way around. After that is said, one needs to be conscious of the fact that signage designers are humans, too, and they cannot be expected to provide perfect systems. These are the key messages that underlie this chapter.

Kirstie Carrick, Peter Pfister, Robert Potter, and Roy Ng (Co-Authors of Chapter 7)

Our interest in aviation human factors and safety has tended to focus on problems that manifest in flight, but the opportunity to research airport signs has brought us back to Earth and to a realization that getting the aircraft from the gate to the runway can be a more complicated and risky activity than one might otherwise imagine. Our approach to the topic was to divide up the proposed sections of the chapter, then research and write up our respective sections — at least, that was the plan. Roy Ng wrote a valuable initial draft of the chapter, concentrating on the ergonomics and technical aspects of airport signage, while Rob Potter researched the ICAO conventions and requirements and collaborated with Kirstie on the introduction. Peter Pfister provided lots of encouragement and good ideas but kept disappearing overseas. Kirstie Carrick researched human factors aspects of signs and interventions for better ground control and rewrote everyone else's contributions to fit her preferred style. Then someone mentioned a case study. Web searches, e-mails, and phone calls resulted in a wealth of information and misinformation, but ultimately in an insight into the difficulties inherent in moving a large aircraft on the ground in poor visibility with inadequate and misleading signs and markings. We hope we have conveyed some of our interest in human factors in this demanding environment.

Donald Kline and Robert Dewar (Co-Authors of Chapter 8)

In writing our chapter, our primary concern was with the visual information needs of older drivers and operators. Despite population "graying" in developed countries, most sign research until recently has been little concerned with the abilities of older drivers and operators. Even now, relatively little is known about how age-related visual change affects sign effectiveness in real-world settings. This is particularly true for transport tasks other than automobile driving such as marine, trucking, air, and rail operations. All too often, as in our chapter, sign effects on the aging must be inferred from what is known about age-related changes on basic visual functions (e.g., reduced acuity, contrast sensitivity and visual fields, eye-movement limitations, susceptibility to glare, slowed visual search, eye diseases, attentional change). For this reason, our review emphasizes understanding age-related changes in the visual system and visual functioning. After introducing the reader to the visual and information-processing limitations of older observers, we attempt to show how transport sign design and use often fail to meet the needs of older drivers and operators. We also show that these, in turn, can affect sign design and evaluation. Finally, we discuss how signs can be optimized for ease of reading and understanding by older persons. Our recommendations are offered with the recognition that changes made to help older persons will benefit drivers and operators of all ages.

Ray Fuller (Author of Chapter 9)

Ray Fuller is firmly of the view that we do not need any kind of new or unique psychological explanation for understanding the motivational and attentional aspects of highway sign use. Thus, in his chapter he has attempted to situate the functionality of traffic signs within an established theoretical framework in psychology, namely, behavior analysis, which provides a robust, functional account of human behavior in diverse settings. This strategy offers the potential advantages of:

- Drawing together disparate empirical observations into a coherent framework (e.g., observations of drivers responding to some sign information without being able to recall the sign; observations of drivers consistently ignoring some signs and so on).
- Explaining why some sign situation combinations do not function as intended (e.g., "Road Work Ahead" sign; School Crossing sign; Speed Limit sign).
- Enabling informed decisions about improved sign design (e.g., School Crossing sign).
- Enabling specification of sign situation design requirements and generation of new signage.

Whether he has succeeded in his broad aim will be up to the reader to decide.

Hashim Al-Madani (Author of Chapter 10 and Chapter 11)

Since the 1940s, researchers in the field of road transport signs have covered a wide range of aspects related to posted signs; however, aspects related to liabilities and psychodemographic characteristics received little attention. In my two chapters an attempt was made to bridge this gap, based on a review of the available studies. The performance of the existing signing system raises serious questions concerning its applicability for various cultures, especially developing countries. Furthermore, the current one-way communication system between the user (the active member) and the sign requires major improvement. Future needs may call for at least two- or three-way systems among the users, sign, and highway control personnel.

Luis Montoro, Antonio Lucas, and María T. Blanch (Co-Authors of Chapter 12 and Chapter 13)

Clearly, the design of Variable Message Signs (VMS) relies on physical, technical, and human factors considerations. "Specific Design Parameters: VMS Part I" focused on providing a useful and handy design toolbox, including the current minimum recommendations and issues for VMS design. Following trends worldwide, two main elements were considered: pictograms and alphanumeric characters ("text"). In both cases, the main design parameters (number of words and rows, message format, abbreviations, message absence, two-frame messages, etc.) were considered in order to offer practical guidelines for design. Experience, however, has taught us that although following this type of recommendation is absolutely necessary, it is not enough for avoiding long-term VMS heterogeneity. Thus, a further recommendation was made advocating consequence-oriented messages.

The main objective for Chapter 13 was to challenge the current levels and progress of ITS/VMS usability in some way. Intelligent Transport Systems offer good potential for increasing road transport efficiency as well as reducing pollution, minimizing congestion, and promoting road safety improvements, but ITS are not without problems. The bond between information and driver, essential for huge and expensive ITS to work, seems somehow overshadowed by normative and technical considerations. For example, whether new and/or transformed road signs take an iconic vs. symbolic form is usually disregarded in spite of its consequences for road users' learning and understanding. Also, dealing with VMS design on a daily basis confirms that VMS devices differ considerably at the local/national and international levels. They not only affect VMS display variability but also distort, change, and extend the "face" of traditional road signs. Last but not least, oncoming presentation systems such as VMS seem to be competing, rather than collaborating, in the task of providing clear and useful information to road users. Nevertheless, we are convinced that under some control, ITS' flexibility and diversity are key issues for future road network improvements.

Michael A. Regan (Author of Chapter 14 and Chapter 15)

I was given flexible marching orders in preparing my chapter (which ended up, on the advice of the editors, as two chapters): write a speculative piece linking Intelligent Transport Systems with traffic signs targeted at a broad audience, not just at traffic engineers, and with a multimodal flavor, as far as possible. This was a challenging exercise because I could find nothing published directly linking the two areas.

For the first chapter, I started by thinking about the attributes of existing road traffic signs (e.g., they are visual, they present static information, and they present information to the driver from outside the vehicle) and then posed a series of questions such as why traffic information should be presented visually and why the information on signs should be static. The idea was then to use these questions to provide structure for a discussion of how emerging developments in traffic technology are changing these attributes of the traditional road sign. I wrote the chapter in this way and ran it past a good friend and respected work colleague, Tom Triggs.

The second chapter, on the human factors implications of alternative means of presenting traffic information using ITS technologies, was also difficult to know how to structure. I started by considering the different sources of traffic information that could impinge on the driver from the various "signs" inside and outside the vehicle and the various deployment scenarios in which these signs would impinge simultaneously on the driver. The biggest decision from there was whether to address the human factors issues relating to each of these deployment scenarios at fairly high levels — which was the approach I took — or to drill deeper and develop a detailed taxonomy of the various options for (and human factors issues associated with) displaying a combined suite of information to the driver from inside and outside the vehicle. I thank Tom Triggs again for a great discussion we had about the form that such a taxonomy might take. I decided, in the end, that it was too ambitious to go down the latter path, which would be the next logical step in developing guidelines for the design of what I call in the chapter the human–machine–environment interface (HMEI) for the display of traffic information. Exposing drivers to multiple sources of information from inside and outside the vehicle has the potential to overload, confuse, and distract them, and I structured the chapter more generally around the key methods that human factors engineers have at their disposal to address these issues: good interface design, procedural guidance and selection, and training. Everything I had written seemed to fit under one of these headings.

The chapters are oriented toward signage in the road transport domain, which I am more familiar with. However, the technologies and human factors issues are relevant to the design of traffic signage in other domains. I have left it up to practitioners in domains outside road transport to assess the relevance of the technologies and issues I have raised to the presentation of traffic information in their domains. My belief is that the issues are much the same across domains.

Author Biographies

This section contains the authors' contact details and short biographies (in their own words). These are presented in order of their contributions to the book. Also, this information is provided in order to offer the reader a greater appreciation of the authors.

Tim Horberry (B.A. [Hons], M.Sc., M.Erg.S., Ph.D.) is a senior research fellow at the Monash University Accident Research Centre (MUARC), Australia. Prior to this, he was a lecturer at the University of Queensland, Australia. His previous jobs have included working for an airline and teaching in a refugee center. Tim is a registered member of the U.K. Ergonomics Society and a former member of the Key Centre for Human Factors and Applied Cognitive Psychology in Australia. Dr. Horberry has conducted a wide range of research projects in the area of transport human factors in the U.K. and Australia; his areas of specialization are the design of visual information, transport safety, driver fatigue, and applied ergonomics. Tim may be reached at Accident Research Centre, Monash University, PO Box 70A, Clayton, Victoria 3800, Australia.

Cándida Castro is a senior lecturer with the Faculty of Psychology, University of Granada, Spain. She holds a Ph.D. in applied experimental psychology. Her areas of specialization are traffic and cognitive psychology, especially attention and perception while driving. Dr. Castro has authored numerous papers in these areas; her Ph.D. thesis was concerned with the arrangement of traffic signs in road environments. She has carried out several overseas working visits, such as at the Psychology Department, Royal Holloway and Bedford New College, University of London, in 1994, where she achieved a postgraduate diploma in European psychology and, more recently, at the Institute of Behavioural Sciences, University of Derby, in 1998. Dr. Castro is a registered member of the ESCOP (European Society of Cognitive Psychology). Cándida's contact information is Departamento de Psicología Experimental y Fisiología del Comportamiento, Facultad de Psicología, Campus de Cartuja, s/n, 18071 Granada, Spain.

Patricie Mertova (Co-Author of Chapter 1 and Chapter 16)

Faculty of Education, Monash University, Victoria 3800, Australia

Patricie is a research fellow at the Faculty of Education, Monash University, Australia. Her background is in the areas of linguistics, translation, communication, and foreign languages. It was this background that played an important part in assisting with the editing of this book.

0-415-31086-5/04/$0.00+$1.50

Francisco Tornay (Co-Author of Chapter 4)

Departamento de Psicología Experimental y Fisiología del Comportamiento, Facultad de Psicología, Campus de Cartuja, s/n, 18071 Granada, Spain

Francisco is a senior lecturer at the University of Granada (Spain). He specializes in research in cognitive psychology, in particular in the field of attention. He is currently involved in basic research, especially in the task-switching paradigm, and in research lines aiming at applying psychological findings to practical problems such as traffic safety and decision making.

Maxwell G. Lay (Author of Chapter 2 and Chapter 3)

RACV, 422 Little Collins St., Melbourne, VIC 3143, Australia

Dr. Lay (Ph.D., M.Eng.Sci., B.C.E., HonFIEAust, FTS, LifeMASCE, MCIArb, FCIT, MAICD) was a principal of Sinclair Knight Merz (1996 to 2003). From 1993 to 1996, he was director of major projects with VicRoads and from 1988 to 1993, he was director of quality and technical resources (responsible for all road design and construction in the State of Victoria). He was executive director of the Australian Road Research Board from 1975 to 1988 and was engineering research manager at BHP from 1968 to 1975. In August 1999, he was named Victorian Professional Engineer of the Year by the Institution of Engineers, Australia. Max has had a road and a park named after him in suburban Adelaide (Maxlay Road) and the annual trophy for the Great Australian Car Rally has been named the Dr. M. G. Lay trophy. He is a member of several professional associations and ISO committees and has won numerous awards and medals. Dr. Lay has authored more than 700 publications on a wide range of topics, including *Source Book for Australian Roads* (three editions), *Handbook of Road Technology* (two volumes, three editions), *History of Australian Roads*; *Ways of the World* (two printings, plus a German edition), and the *Encyclopaedia Britannica* entry on roads. He has recently published a history of metropolitan Melbourne's streets and roads, *Melbourne Miles*.

Terry Lansdown (Author of Chapter 5)

Centre for Occupational Health and Safety, Heriot–Watt University, Edinburgh, EH14 4AS, U.K.

For more than 10 years, Dr. Lansdown has been researching transportation human factors. He is employed as a lecturer at Heriot–Watt University. Dr. Lansdown contributes to several vehicle ergonomics BSi committees and has managed and participated in numerous U.K. and European research projects. His research interests include: visual attentional allocation/distraction, human error, and system safety and usability.

Alexander Borodin (Author of Chapter 6)

Manager Human Factors, Risk Unit, Queensland Rail, Brisbane, Australia

Alex holds a degree in microelectronic engineering from Griffith University. His specialty is computer systems engineering. He has worked for Queensland Rail since 1995, having started as a software and systems engineer in the Signal and Operational Systems section — a post that he held for 4 years. Following this, he moved to Network Access Group where he was the signal systems engineer for 2 years. Currently, Alex holds the position of manager of Human Factors in QR's risk unit, a post responsible for the management of the SPAD (signals passed at danger) and Derailment Reduction Programs as well as the QR Human Factors Strategy. His areas of interest are human factors in systems design and safety, methods for developing safety cultures, and scanning the horizon for what is being achieved in risk management in various industrial organizations worldwide.

Peter Pfister (Co-Author of Chapter 7)

Associate Professor, The University of Newcastle, Callaghan, NSW 2308, Australia

Peter holds a diploma in physiology and pharmacology from Switzerland and a B.A. in psychology and biology from Macquarie University, Australia. He completed a Ph.D. at Newcastle, Australia, studying stress with Dr. Al Ivinskis. His research focus is in the area of human factors, especially as it relates to aviation and other high-tech industries. Generically, he is interested in: improved performance, conservation of resources, adaptation to and adoption of change, and promotion of safety, especially in the operational environment.

Kirstie Carrick (Co-Author of Chapter 7)

The University of Newcastle, Callaghan, NSW 2308, Australia

Kirstie has a B.S. (Hons) and M.S. from the Australian National University. She is currently studying for a Ph.D. in the area of memory and cognition. She is a human factors psychologist specializing in aviation industry issues, especially human factors training and safety. Kirstie is currently the convenor of aviation postgraduate programs at the University of Newcastle, Australia.

Robert Potter (Co-Author of Chapter 7)

Aviation Performance Systems Pty. Ltd., Australia

Rob, CEO of Aviation Performance Systems Pty. Ltd., holds an M.Sc.Stud. from the University of Newcastle, Australia, and an airline transport pilot's license

(ATPL) with approximately 8000 hours of airline flying. Rob's consulting work is focused primarily in providing assistance to regional aircraft operators in the area of aircraft performance and operations. He is currently studying for a Ph.D. in this field.

Roy Ng (Co-Author of Chapter 7)

Airplan, Sydney, Australia

Roy has more han 30 years' experience in engineering consultancy, with prime responsibility for the delivery of major infrastructure projects, from concept development to construction and commissioning. He is currently an associate director of Airplan, a multidisciplinary firm that specializes in airport and aviation business consultancy work around the globe. Roy has been closely associated with facility planning/development, strategic capacity assessment, and land use planning at Sydney Airport over the past 14 years. He has also been responsible for other airport projects at Abu Dhabi and at the new international airports at Hong Kong and Kuala Lumpur.

Donald Kline (Co-Author of Chapter 8)

University of Calgary, Calgary, Alberta, Canada

Donald is a professor in the Departments of Psychology (PACE) and Surgery (Ophthalmology) at the University of Calgary and the director of the Vision and Aging Laboratory. His research interests include the neural and optical mechanisms of visual aging, visual health eye surgery, visual function and quality of life, and the effects of visual change on tasks such as driving, reading, and the use of optical and image-processing techniques for optimizing font and sign legibility. Information about research and teaching activities in the Vision and Aging lab is available on the lab's Web site, http://www.psych.ucalgary.ca/PACE/VA-Lab/.

Robert Dewar (Co-Author of Chapter 8)

University of Calgary, Calgary, Alberta, Canada

Bob is professor emeritus of psychology at the University of Calgary, where he has been teaching since 1965 and is now president of his own consulting firm, Western Ergonomics, Inc., specializing in visual information systems, driver behavior, and traffic safety.

He has written a chapter, "Road Users," for the fifth edition of the *Traffic Engineering Handbook* and a book, with Paul Olson of the University of Michigan, *Human Factors in Traffic Safety*. His consulting work on traffic signs includes design of symbols, changeable message signs, bilingual freeway signs, evaluation of tourist direction signs, and methods for the evaluation of signs. He has presented several workshops on human factors in traffic safety for traffic engineers, police, driver educators, and traffic safety researchers, as well as presented (with others) a series of workshops on traffic safety in work zones for the Transportation Association of Canada. Dr. Dewar was a member of the consulting team responsible for rewriting the Canadian Manual on Uniform Traffic Control Devices and of the team writing two chapters of the Ontario Traffic Manual.

Ray Fuller (Author of Chapter 9)

Department of Psychology, Trinity College, Dublin 2, Ireland

Ray is a former head of the Department of Psychology, Trinity College, Dublin, and also a former president of the Psychological Society of Ireland. He is currently a member of two working parties of the European Transport Research Council (Road User Behavior and Traffic Regulation Enforcement). For the last 25 years his research has focused mainly on transport safety, with particular emphasis on understanding driver behavior and the contributing factors underlying collisions on the roadway. In the last decade or so, he has extended this research to investigation of human factors in aviation safety. He has edited and part-authored seven books and published well over 100 articles in the international research literature. His recent major works include *Human Factors for Highway Engineers* (with Jorge Santos) and an evaluation of Operation Lifesaver (a high-enforcement safety intervention) for the National Roads Authority of Ireland. He has also recently completed a monograph, Education and Training of the Driver in Post-Primary Schools, for the National Council for Curriculum and Assessment at the behest of the Minister for Education and Science.

Hashim Al-Madani (Author of Chapter 10 and Chapter 11)

Associate Professor, Department of Civil and Architectural Engineering, University of Bahrain, P.O. Box 32038, Bahrain

Hashim (B.S., M.S., Ph.D.) is associate professor, of transportation, Department of Civil and Architectural Engineering, University of Bahrain. He has published more than 25 papers in refereed international journals, conferences, and books. He is the founder and a board member of the Center for Transport and Road Studies, College of Engineering, University of Bahrain. Dr Al-Madani is a research committee member of the General Committee for Road Safety and for the Society of Road Accident Victims, Bahrain. He is the chairman of the Organizing Committee of Safety on Roads: International Conferences — SORIC and is the co-chairman of the organizing committee for Transport and Road Symposiums.

Luis Montoro (Co-Author of Chapter 12 and Chapter 13)

INTRAS, Instituto del Tráfico y Seguridad Vial, Universidad de Valencia, Hugo de Moncada, 4-bajo, 46010 Valencia, Spain

Luís, professor of road safety at the University of Valencia (Spain) and director of the Traffic and Road Safety University Institute (INTRAS), has written numerous road safety books, papers, and technical reports and has become a road safety expert for the Spanish media. He is currently involved in several different national and international research projects and is a member of the executive committees of different road safety associations. He has earned different national and international honors and awards recognizing his influential research work (International Bridge Tunnel and Turnpike Association research award; Road Safety merit medal conferred by the Spanish Road Directorate — DGT).

Antonio Lucas (Co-Author of Chapter 12 and Chapter 13)

INTRAS, Instituto del Tráfico y Seguridad Vial, Universidad de Valencia, Hugo de Moncada, 4-bajo, 46010 Valencia, Spain

Antonio is a technical researcher at INTRAS (University of Valencia). His main research interests are new road information technologies and social norms in road traffic; these are currently being developed within the framework of the ITS program jointly developed by the Spanish DGT and INTRAS. He is also involved in Euroregional ITS projects (European Commission), representing DGT in the European VMS Platform (WERD/DERD).

María T. Blanch (Co-Author of Chapter 12)

Universidad de Valencia, Hugo de Moncada, 4-bajo, 46010 Valencia, Spain

María Teresa is currently undertaking doctoral studies on human activity and psychological processes. She holds an official grant (National Training Program for University Teachers, General University Directorate) ascribed to the University of Valencia.

Michael A. Regan (Author of Chapter 14 and Chapter 15)

Accident Research Centre, Monash University, P.O. Box 70A, Clayton, Victoria 3800, Australia

Michael is an applied experimental psychologist with B.S. (Hons) and Ph.D. degrees in human factors and ergonomics from the Australian National University. Since 1997 he has worked as a senior research fellow at the Monash University Accident Research Centre (MUARC) in Melbourne, Australia, specializing in research on behavioral and ergonomic factors that contribute to road crashes. Prior to that he was the manager of road user behavior at VicRoads, where he was responsible for developing policies and programs for reducing road trauma attributable to human factors in the Australian State of Victoria. Before that, Dr. Regan worked as a research scientist with the Australian Defense, Science and Technology Organization (DSTO), specializing in the ergonomic design of military aircraft cockpits. Michael's current research interests are in the areas of Intelligent Transport Systems (ITS), driver distraction, HMI design for intelligent vehicles, human-in-the-loop driving simulation, and driver training and licensing.

He is the project director of the internationally known TAC SafeCar project, an on-road study of long-term driver adaptation to multiple in-vehicle Intelligent Transport Systems. In 2002, Dr. Regan was appointed as the Australian member of the International Organization for Standardization (ISO) Technical Committee 22, Subcommittee 13, which develops international standards for the ergonomic design of cockpits in road vehicles. Michael is the author or co-author of more than 90 published reports, articles, and conference papers.

Index

A

A5 page size, standard for airport diagrams, 96
Abbreviations
acceptable uses in VMS systems by culture, 192
in VMS systems, 190
Accident analysis, 109–110
Accident involvement
with drivers at fault, 178
unrelated to sign comprehension, 177
Accident rates
among aging drivers, 116
among drivers with binocular visual field losses, 119
increased through information overload, 202
increasing steadily throughout 20th century, 200
Accident warnings, pictograms for VMS systems, 207
Accidents, due to poor signing, 51
Accommodation
age-related decline in, 117
head-up displays to avoid shifts in, 233
Accuracy, of transport signs, 6
Acquisition of information, time required for, 8
Acting distances, 43
Action potential
requirement for driver response determining, 149
of signs, 54
Adaptive cruise control, 220
Advance direction signs, 45
Advanced driver assistance systems, 220
Advertising signs, roadside, 34
proximity to traffic signs, 63
Advisory signs
in on-board intelligent speed adaptation systems, 217
in Variable Message Signs (VMS), 185
Aerodrome chart sample, 97
African drivers, and sign comprehension, 170
After-landing checks, 103
Against-the-rule astigmatism, 117
Age, 6
and advantages of augmented signage technologies, 228
compensatory behavioral modification associated with, 73
and experience correlated with sign comprehension, 175
fixation durations and, 51
in scenes high in clutter, 150
in relation to driver gender, 173
and risk of ignoring warning signs, 175
sign comprehension correlated with, 172, 174–176
and technologies for optimizing traffic environment, 229
and visual acuity problems, 10, 73, 116
Age-related compensatory strategies, 130
FHA older driver highway design handbook, 229
Age-related macular degeneration (ARMD), 116
Age-related visual decline, 116–117
accommodation, 117
degraded color discrimination, 118
discrimination distance and, 122
dynamic vision, 118–119
reduced retinal illumination, 117–118
refractive problems, 117
saccadic eye movements, 119–120
smooth pursuit eye movements, 119–120
static acuity, 118
static contrast sensitivity, 118
and strategies for enhancing signage, 128–129
susceptibility to transient glare, 117
useful field of vision, 119
visual search tasks, 120
Aigrain Report, 200
Air traffic control
automated alerts for, 106–107
causing runway incursions, 105
controlling runway traffic permissions, 98
datalink information, 106
Aircraft
control of, as distraction from signage, 103
separation via automated alert systems, 107
Aircraft location systems, 106–108
Airport diagrams, 96
impact of inaccuracy
on runway incursions/accidents, 108–111
on situational awareness, 103
Airport expansions, out-of-date on airport diagrams, 103

Airport layout and complexity, as factor in runway incursions, 105
Airport lighting, as factor in runway incursions, 105
Airport Movement Area Safety System (AMASS), 106–107
Airport signing
case study, 108–111
environmental influences on visibility of, 102–103
as factor in runway incursions, 105
interventions against runway incursions, 105–108
rationale for, 95–97
role of International Civil Aviation Organization (ICAO) in, 98
sign locations, 101
size specifications, 101–102
standards for, 98–100
technical difficulties involving, 102
Airport surface detection equipment—model X (ASDE-X), 106
for smaller airports without AMASS capability, 107
Airservices Australia
definition of runway incursions, 104
runwy incursion task force, 112
Al-Madani, Hashim, 244, 251
Alcohol detection system, 220, 221
Alcohol ignition interlock features, 221
Alertness
of aging operators, 121
driver, 37
Algorithmic approach to road sign interrpetation, 205
Alliance of Automobile Manufacturers (U.S.), design guidelines for human-machine interfaces, 230
Alphabetic signs, 9
Alternating messages, in VMS systems, 190, 194
Ambiguity, 39
American Association of State Highway Officials, and U.S. national route numbering, 18
American National Standards Institute (ANSI), 160
American signing manuals, historical development of, 21
AMITCS traffic management system (Japan), 200
Angle of observation, and sign conspicuity, 121
Annex 14: Standards and Recommended Practices for Aerodromes, 98
Antecedent events
in contingency model of behavior, 136–137
examples of, 137
Approach speed, determining sign size characteristics, 148
Appropriate action, 40
Arab region drivers
Lane Reduction signs comprehended by, 170
poor comprehension of signs, 157, 170
signs best comprehended by, 160
Slippery When Wet signs comprehended by, 170
Armagh accident of 1889, 83
Arport signing, historical development of, 95–96
Aspect sequencing, in railway signage, 85
Assessment environments, 78
Assessment methodologies, 78
Astigmatism, 117
At-grade intersections, problematic for older drivers, 229
ATC datalink information, installed in cockpit displays, 106
Attention, 240
ability of VMS to attract driver, 226
information overload discouraging, 202
seven aspects of, 149
Attention conspicuity, 33
and brightness contrast between sign and its background, 149
Attention signs, historical development of, 20
Attention value, 149
dynamic variation in auditory signals and, 150
Attentional capabilities, human, 73
Attentional spotlights, 79
Auditory alerts and signals
advantages of information presented as, 228
driver acceptance of, 233
enhancing attention and compliance, 150–151
in intelligent speed adaptation (ISA) systems, 217
Augmented signage
issues and disadvantages of, 229–230
net effect on driver workload and distraction, 231
potential advantages of, 228–229
timing issues related to, 231
Australia, example of unusual warning sign, 5
Australian Standard Rules for the Design, Location, Erection, and Use of Road Signs and Signals, 163
Auto-dimming systems, 129
Automated alerts
for air traffic controllers, 106
not provided directly to aircraft, 107–108
Automatic train protection systems, 92
lack of standardization in Europe, 93
Aviation signing, 13, 242
application of VMS to, 186

development of pilot associates in, 232

B

Background complexity, as determinant of sign conspicuity, 56
Background luminance, affecting legibility distance, 121
Behavior
in contingency model, 136–137
goal oriented and intentional, 137
Behavior analysis, 136, 243
Behavior-response consequences, role of directional signs in resolving uncertainty about, 148
Behavior theory, 136
Behavioral plasticity, 136
Behavioral testing, of warning signs, 7
Bicycle clubs, and historical origins of warning signs, 18
Bilingual messages, in VMS systems, 191–192
Black-and-white signs, 45
legibility distances for, 148
Black boxes, aiding in speed limit enforcement, 144
Blanch, Maria T., 244, 252
BLOCK LIMIT boards, 86, 88
Blooming phenomenon, 78
Blue-green color spectrum, age-related decline in sensitivity to, 118
Blue signs, 45
Borders, and perceived urgency, 62
Borodin, Alexander, 242, 249
Braking
potential for premature with use of emerging technologies, 228
response time, 41
and safety liabilities in human-machine interface design, 231
Breaches of duty, 164
Breakaway poles, 46
Brightness. *See also* Luminance
affecting legibility distance, 121
and sign conspicuity, 121
British Rail route signaling principles, 85
Brown signs, 45
Bureau of Public Roads, 36
Buried inductive loops, in vehicle-activated signs, 216

C

Cardinal Richelieu, 18
Carrick, Kirstie, 242, 249
Castro, Candida, 240, 247
Cataracts, 116
Categories, of runway incursions, 104
Center line lights
on airport runways and taxiways, 103
in analysis of Singapore runway incursion accident, 110
Changeable message signs, 29
Character size, affecting legibility distance, 121
Characters per row
effect on message lengths in VMS systems, 190
in VMS systems, 188–189
Charles de Gaulle Airport, use of towable barriers, 110
Chicanes, as traffic calming measure, 145
Children Crossing sign, 143
showing a reliable contingency, 143
as unreliable contingency, 141–142
Chromaticity, 42
CIE chromaticity diagram, 42
Circular signs, for instructional messages, 7
Clearview font, optimized legibility of, 126
Clutter
in airport diagrams, 96
avoiding in signage design for older drivers, 129
effects on older drivers, 121
engineer strategies for reducing, 151
eye fixations increased in presence of visual, 150
Cockpit displays
coding scheme issues in transition from analog to CRT displays, 227
guidelines for designers of, 230
head-up displays, 108
incorporating global positioning systems (GPS) and ATC datalink information, 106, 108
Coefficient of luminance intensity (CIL), 27, 28
Cognitive distance, 42
Cognitive load. *See also* Workload issues
on drivers, 91
pushing pilots and air traffic controllers to limits of capacity, 112
Cognitive response, 40
Color combinations
affecting detection of runway hold position signs, 106
avoiding color-blindness problems, 78
for mandatory instruction signs in airports, 98
for nighttime runway and taxiway lighting in airports, 103

for runway/taxiway information signs in airports, 99
of VMS *vs.* traditional traffic sign information, 227
white on blue, 20
Color contrast
and sign conspicuity to aging drivers, 121
between signs and advertisements, 62
Color discrimination, age-related decline in, 118
Color inversion
adoption after Vienna Convention, 207
in VMS systems, 206
Colors
available in VMS systems, 206
effectiveness increased with positional cues, 78
meanings used on signs, 43
and sign conspicuity, 121
signaling by, 22
and signification in railway signage, 85
stereotypes associated with specific, 78
Compass directions, identifying runways by, 96
Compatibility
of design and timing of internal/external information, 230
of information design betwen VMS and conventional traffic signs, 227
of messages from inside and outside vehicles, 231
Compliance issues, 6, 241
driver motivation and, 135–136
effect of repeating speed signs at intervals, 145
with ICAO standards for airport signage, 101
police presence and, 144
psychology of, 136–147
speed limit signs in VMS, 203
Comprehensibility, 38–40
evaluating current sign systems in terms of, 155
as requirement of transport signs, 5
of signs, 4
studies on, 156
techniques for evaluating and measuring, 157–161
Comprehension
age-related problems of, 125
American studies, 160, 162
Arab world studies, 160, 161
Canadian studies, 160, 162
correlation with driver age, 171, 174–176, 175
correlation with driver education, 171–172, 176–177
correlation with driver experience, 172, 174–176
correlation with driver gender, 172–174
correlation with driver income, 176–177
correlation with driver socioeconomic and safety characteristics, 179–181
correlation with seat belt usage, 179
and driver origin, 170–172
of fixed road signs *vs.* Variable Message Signs, 201
Idaho-based studies, 160
by minorities and non-native speakers of English, 170
Texas-based studies, 160
three factors most strongly affecting, 179
unrelated to accident involvement, 177
unrelated to citations, 179
unrelated to marital status, 172
VMS issues related to, 186
Comprehension testing, 10
Congestion warnings, 74
lack of VMS pictograms for, 207
in VMS systems, 187
Japan, 204
Consequences, in contingency model of behavior, 136–137
Conspicuity, 33–34, 73
and needs of aging drivers, 121
and needs regarding, 76
relative, of safety-relevant signs, 151
as requirement of transport signs, 4
Construction zones, problematic for older drivers, 229
Contingency model of behavior, 136–137
conditional rules in, 147
uncertain contingencies on unfamiliar roads, 146–147
unreliable contingencies undermining sign effectiveness, 141–145
Contour interaction susceptibility, 38
and sign borders, 62
Contrast
as determinant of sign conspicuity, 56, 121
heightening in signage design, 129
low between sign and background, 120
minimizing older drivers' acuity limitations, 118
Contrast sensitivity
and discrimination distance, 122
losses in aging drivers, 116
Contrast sensitivity function (CSF), 118
Control level of driver behavior, role of VMS in, 186
Control tower, runway signs requiring authorization from, 98
Convention on Road Signs and Signals, 2, 27
Cooperative ITS technologies, 215
Japanese initiatives, 221

in traffic information systems, 219–220
Correct action, 40
Costs and benefits
determining driver acceptance of in-vehicle technologies, 233–234
of high-technology air safety interventions, 112
of traffic sign improvements *vs.* traffic channelization improvements, 177
of traffic signs, 72–73
of updating digital road maps and global positioning systems, 218
of vehicle-activated signs, 216
Credibility
of transport signs, 6, 54
of VMS messages, 202
Credible action, 40
Cross-cultural uniformity, 155–165
Cross Road symbol, legibility distance of, 123
Curbing, 26
Current sign systems, concerns about applicability of, 165

D

Danger warning signs, 2
red color of, 22
Derailments, avoiding by indicating point switches and maximum allowable speeds, 83
Design variables, 6, 13
in human-machine interface design, 230–232
in VMS systems, 206–209
use of standard *vs.* innovative coding schemes, 227
Destination signs, high priority for attention, 149
Detail, detection of, 34–36
Dewar, Robert, 243, 250–251
Diabetic retinopathy, 116
Diagonal cross-shaped signs, 44
Diagrammatic signs, 45
Diamond-shaped warning signs, 163
Digital audio broadcasting (DAB), 219
Digital road maps
in in-vehicle navigation systems (IVNS), 222
incorrect identification of side road speed limits in, 218
in intelligent speed adaptation (ISA), 217
NextMAP project for increasing accuracy of, 218
no longer required on-board with emerging technologies, 222
updating issues, 218
Direct instructions, written on roadway, 147
Direct traffic control (DTC) system, 86, 88
Direction sign distances, 42, 135
Direction signs, 44–46
as discriminative stimuli for route choices, 139, 140
information overload of, 8
and resolution of uncertainty about behavior-response consequences, 148
Directions for English Travillers, 17
Dirver motivation, 149
Disability glare, age-related problems with, 73
Disc-shaped signs, 44
Discrimination distance, 122
Discriminative stimuli, 136
directional signs as, 139
dirrect instructions on road surface, 147
fabric mockups, 145–146
penalty warning increasing compliance, 144
requiring reliable contingencies, 141–145
speed limit signs as, 142–144
traffic signs as, 138–140
virtual, 146
Display location, in in-vehicle systems, 232–233
Disregarding signs, 201–203
Distance markers
on airport runways and taxiways, 101
history of, 17
Distance to event, in VMS systems, 195
Distractions, 63–65
affecting visual search tasks in aging drivers, 120
avoiding in augmented signage design, 229
encountered by air crew during taxiing, 103
from in-vehicle technology, 74
from inconvenient display locations of in-vehicle technologies, 232–233
from intelligent transport systems, 223
net effect of augmented signage on, 231
potential with multiple sources of traffic informatoin, 229–230
training users in avoiding via in-vehicle technologies, 235
by VMS display of unimportant messages, 193
Diversion signs, new pictograms needed for VMS systems, 205
Divided Highway, comprehensibility by age, 125
Dominant-subordinate, large-small delay, 60–61
Dot matrix technology, for variable message signs, 29
Drive traffic management system, 200
Driver alertness, 37
Driver at-fault accident involvement, unrelated to sign comprehension, 178
Driver attentional allocation, 71–80

Driver behavior, classification of, 186
Driver education, unrelated to knowledge of traffic signs, 177
Driver expectations, 149
Driver experience
and age correlation with sign comprehension, 175
correlation with sign comprehension, 172, 175–176
in relation to gender, 173
Driver line of sight, 31
Driver motivation, 6, 53–56
Driver origin, sign comprehension in relation to, 170–172
Driver priorities, 76
Driver recall, 50–51
Driver reminder devices, 91
Driver responses, to traffic signs, 53
Driver visual information acquisition strategies, 79
Driving awareness, factors influencing, 54
Driving experience
affecting awareness of signs, 54
correlated with sign recognition, 141
Drowsiness detection system, 220
Duc de Sully, 18
Dynamic contrast sensitivity, 118–119
Dynamic navigation systems, 74
Dynamic speed indicators (DSIs), 88, 89
Dynamic variation, 150
Dynamic vision, age-related decline in, 118–119
Dynamic visual acuity, 118

E

E-mail, in-vehicle technology for, 74
Ecological validity issues, 7
Edge contrast, as determinant of conspicuity, 33
Education, correlation with sign comprehension, 171–172, 176–177
Educational kits, for general aviation pilots, 106
Effectiveness of signage, 74–75
for older drivers, 157
strategies for enhancing
bordering signs, 62–63
new signs, 56–57
reducing visual clutter, 63–65
repeating signs, 57–59
signs on same post, 60–61
Effectiveness rates, 8
combining text with symbols, 11
at localized/behavioral level, 13
nighttime *vs.* daytime, 12
of signs, 30
of Variable Message Signs, 72
Electronic ballasts, for airport runway fluorescent lighting systems, 102
Emerging technologies
Intelligent Transport Systems (ITS), 214–215
variable message signs (VMS), 185–197, 199–210
EMILY project, 218
English-Welsh bilingual messages messages, 192
Environmental attentional demands, 77
Environmental factors
affecting legibility, 121
in sign conspicuity, 55
Equilateral triangle shape, 44
Ergonomic problems
of designing future augmented signage, 230
of runway/taxiway signage, 100
ERTICO-ITS Europe, 218
traffic message channel initiative in driver's preferred language, 219
Euroface font, 77
European Action Plan for the Prevention of Runway Incursions, 106
European Commission
projects for improving global positioning systems, 218
Statement of Principles for design of human-machine interface, 230
European Countries-Cooperation in the Field of Scientific and Technical Research (EUCO-COST), 200
new pictograms proposed at, 205
European ERTMS standard, 92–93
European Rules Concerning Road and Traffic Signs and Signals, 162
European Transport Safety Council, 200
Experience-per-accident ratio, 178–179
Exposure rate. *See* Driver experience
Eye diseases, effect on aging and visual decline, 116
Eye fixations
in aging drivers, 121
in high-clutter areas, 150
Eye glance measurement, 78
Eye movement analysis, 79
in older drivers, 119–120
Eye movement process, 34, 51
recording in measurement of sign effectiveness, 50

F

Familiarity

as determinant of driver awareness and sign conspicuity, 57
as determinant of legibility, 121
Fatal accidents
decreased by improved traffic signs at intersections, 177
increased with density of traffic flows worldwide, 200
Features, identifying relevant, 26
Federal Aviation Administration (FAA)
advisory circulars specifying sign systems and standards, 99
airport movement area safety system (AMASS), 106–108
definition of runway incursions, 104
Federal Highway Administration, 155, 162
development of older driver highway design handbook, 229
Female drivers, sign comprehension by, 172–174
Fiber optics, for variable message signs, 29
Filtering information
and ergonomics of human-machine interface design, 234
in future augmented signage technology scenarios, 228–229
role of automated workload managers in, 232
Finger posts, 18
Fixation durations, longer at nighttime, 51, 54
Fixed board signals (railway), 86–88
Flags, signaling by, 22
Flashing lights, signaling intensity, 150
Flight deck communications, as distraction from airport signage, 103
Flight Safety Australia, 112
Fog warnings, new pictograms needed for VMS systems, 205
Following distance
vehicle-activated signs related to, 216
warning systems, 215, 220
Fonts
Clearview, 126
design issues for future VMS systems, 227
Euroface, 77
Helvetica, 77
and legibility issues, 77
legibility variations in, 126
lower-case for unimportant VMS messages, 93
modified Series E letterset, 36
Forward collision warnings, 220
Four-aspect signals, 86
Four car blinkers, as extension of variable road signs, 209–210
Fourier analytic techniques, in symbol design, 127–128
Frangibility
of airport runway and taxiway signs, 101
of signpoles, 46
Franklin, Benjamin, 18
Freeway instructions, provided by on-board ITS systems, 221
French Touring Club, 18
Fuller, Ray, 243, 251
Functional effectiveness, of drivers, 79
Functional road signs, 73

G

Galileo project, 218
Gender
correlation with sign comprehension, 172–174
in relation to driver age, 173
in relation to experience, 173
General aviation (GA) flights, runway incursions made in Australia by, 105
Geometric requirements, of signs, 30–31
German FFB and LZB systems, 93
Glance legibility, 37–38
variations by age group, 124–125124–125
Glare
age-related susceptibility to, 117, 124
in VMS systems, 206
Glaucoma, 117
Global positioning system (GPS)
in cockpit displays, 106, 108
in intelligent speed adaptation (ISA), 217
speed limits for side roads incorrectly identified in, 218
as vehicle-based technology, 215
Goal-oriented behavior, 137
Goals, of driving, 128
Golden milestone, 17
Grade separation intersections, problematic for older drivers, 229
Green, maritime symbolism of, 22
Green signs, 45
GSM telephones, in emerging IVNS technologies, 222
Guidance level of driver behavior, role of VMS in, 186
Guide signs and information, 26, 214
emerging technologies in, 221–223
trip planning asisstance, 223

H

Hazard control, hierarchy of, 7–8

Hazards, from visual overload of signage, 76
Head-up displays
in aircraft cockpits, 108
in in-vehicle traffic information technology systems, 233
Headlamp reflection
designing for low-glare, 129
transient glare from, 117
Headway distance, VMS solutions to compliance with, 203
Helvetica font, 77
High-density discharge (HID) headlights, 78
High-technology interventions, in airway safety, 106–108
Hill symbol, comprehensibility by age, 125
Horberry, Tim, 240–241, 247
Human error, role in transport accidents, 49
Human factors, 6, 14
in airport signing, 100–104
overriding technological interventions in airway safety, 112
Human-machine-environment interface (HMEI), 230
Human-machine interface design, 241, 245
development of workload managers, 232
display location issues in, 232–233
documentation and user training issues in, 234
driver/user acceptance issues, 233–234
guidelines and standards for, 230–232
Humps, as traffic calming measure, 145
Hyperopia, 117

I

ICAO signage conventions, lack of pilot education in, 100–101
Icons
difficulties in developing new pictograms for VMS systems, 205
vs. symbols in VMS systems, 203
Illumination
color discrimination reduced with poor, 118
degraded at nighttime for older drivers, 123–124
designing signs for low glare, 129
specifications for airport runways and taxiways, 101
and visual problems of aging drivers, 116, 117
Image processing techniques, use in optimizing signs, 129
Imperial Rome, milestones in, 12, 17
In-cab signaling, 92
In-ground inductance loops, 107
In-vehicle navigation systems (IVNS)
emerging technologies in, 222
enhancing mobility of older drivers, 116
future advantages of, 228
global positioning system for transmitting speed limits, 151
as one among multiple sources of traffic information, 226
pocket PCs as, 232
presenting problems to VMS designers, 206
In-vehicle signage technology, 74
age-related difficulties in reading, 116
distractional capabilities of, 74
as one of multiple sources of future traffic information, 227–230
Inaccurate information, on VMS systems, 202
Income, correlation with sign comprehension, 176–177
Incongruent messages, 54
Indication signs, 8
identified more quickly and accurately if worded, 60
vs. warning signs, 7–8
Inexperienced drivers, higher level of sign recall, 54
Information overload, 8, 241
avoiding in augmented signage design, 229
and effect on aging driver comprehension of transport signs, 120
in HMI design considerations, 231
potential with multiple sources of traffic information, 229–230
in VMS systems, 201–203
Information-processing limitations, 78, 242
considerations for multiple sources of traffic information, 230
Information redundancy, 54
Information signs, 26
on airport runways and taxiways, 99
Information technology
enhancement of road traffic with, 200
problems with continuous innovation of VMS systems, 206–207
Informative rerouting messages, in VMS systems, 194
Informative signs, 26
basic message format for VMS systems, 194–195
Infrastructure-based ITS technologies, 215
Infrastructures, optimizing existing *vs.* road system expansion, 200
Innovation, *vs.* standardization in VMS systems, 206–207
INRETS, 50
Inset edge lights
on airport runways and taxiways, 103

findings in Singapore accident analysis, 109, 110
Institute for Transport Studies, 50
Intelligent speed adaptation (ISA) systems, 217–218, 220
and driver loss of control issues, 236
Intelligent Transport Systems, 4, 13, 214–215, 237, 241, 244, 245
advanced driver assistance systems as, 220–221
effectiveness dependent on design of road signs, 201
functions of, 185
future advantages of, 228
as one of multiple sources of traffic information, 226
potential for driver confusion, overload, distraction, 223
traffic management and control as goals of, 199
Interactive signs, vehicle-activated signs (VAS), 215–216
Interletter spacing, 36
Internal transillumination, 32
International Civil Aviation Organization (ICAO)
position on runway incursions, 104
role in airport signing, 98
International Organization for Standardization (ISO), 160
design guidelines for human-machine interfaces, 230
International traffic signs, 2–4
history of development of, 20
need for internationally valid pictogram set for VMS systems, 196
uniformity and differences in, 161–164
Internationalization, 240. *See also* Standardization issues
of road sign pictograms, 203–204
vs. proliferation of new local/national pictograms, 207–209
Intersections, improvement in trafic signs reducing accidents at, 177
Irrelevance
of VMS messages, 202
of warning signs with unreliable contingencies, 141–142
Italian Touring Club, 18

J

JAA Strategic Safety Initiative, 106
Jeppesen airport diagrams, 103, 108–109
Junction route indicators (JRIs), 89
Junctions, complexity of, 76

K

Kline, Donald, 243, 250

L

Laboratory studies, of traffic sign effectiveness, 50
Landolt C acuity target, 122
Lane change collision warnings, 220
Lane closures, 187
format order for text messages in VMS systems, 191
Lane departure warnings, 220
Lane Ends symbol sign, comprehension of, 170
Lane narrowing, as traffic calming measure, 145
Lane Reduction signs, comprehended by Arab *vs.* U.S. drivers, 170
Lanes Closed message element, in VMS systems, 191
Language proficiency
as factor in runway incursions, 105
in iconizing of road signs on VMS displays, 203
Lansdown, Terry, 242, 248
Laser lights, at runway hold points, 106
Lay, Maxwell G., 241, 248
League of Nations Convention 1931, 19
League of Tourist Associations, 20
Learning psychology, 136–147
Legal blindness, 116
Legal issues, in conventional sign obsolescence, 236
Legends, 29
Legibility, 34
affected by age-related visual decline, 116
in airport signage, 101
establishment of minimum distances for, 129
and font selection, 77, 126
issues for older drivers, 121–123
optimizing
symbol signs, 126–128
text signs, 126
as requirement of transport signs, 4, 6
sign borders and, 62–63
of symbolic *vs.* text-based signs, 9–11, 30
for older drivers, 123–124
Legibility design, transcending lighting conditions, 124
Legibility distance, 35, 36, 42, 121
and approach speeds, 148

of retroreflective signs, 28
in VMS systems, 190
of warning sign symbols, 123
Legibility time, 37–38
Legislation and liabilities, international variations in, 164–165
Letter size guidelines, 122
Lettering size, 38
Lifetime accident involvement, unrelated to sign comprehension, 177–178
Light/dark differences, age-related ability to discriminate, 118
Lighted taxiway signs, 102
Lighting uniformity, on airport runways and taxiways, 101
LIMIT OF SHUNT boards, 86
Link messages, in VMS systems, 194
Local Government Board (Britain), 20
Location
of airport signs relative to edge of taxiway, 101
and sign attention value, 149
Location markers
on airport runways and taxiways, 99
railway, 90
weather affecting visibility of, 102–103
Low-glare headlights, 129
Lower-case fonts, for unimportant messages in VMS systems, 193
Lucas, Antonio, 244, 252
Luminance contrast ratio, 32
affecting legibility distance, 121
Luminance contrast steps, use of large to enhance legibility, 129
Luminosity, 31
applicability to VMS design, 186
as determinant of sign conspicuity, 55
effects on older drivers, 121, 122
Luminous flux, 31
Luminous intensity, 31

M

Male drivers, sign comprehension by, 172–174
Mandatory instruction signs, 135
in airport signing, 98
Mandatory signs, 3, 26
Mandatory speed limitation, via intelligent speed adaptation (ISA) systems, 217
Manual on Uniform Traffic Control Devices, 3, 19, 21, 60, 162, 163
Maps, enhancing mobility of older drivers, 116
Marital status, unrelated to sign comprehension, 172–174
Maritime influences, on history of signs, 21–22
Maritime signing, application of VMS to, 186
Mechanically interlocked signaling systems, 83
Medians, 26
Memory of signs, 53
Men Working symbol, comprehensibility by age, 125
Mental workload assessment, 78
Mertova, Patricie, 247
Message absence, in VMS systems, 192–193
Message content, in VMS systems, 189
Message length, in VMS systems, 190
Message prioritization, in VMS systems, 188
Michelin, Andre, 18
Microwave detector heads, in vehicle-activated signs, 216
Mileposts. *See* Milestones
Milestones
historical use of, 12
in Roman times, 17
Minimum angle of resolution, 35
Minimum distance, 6
Minimum required visibility distance (MRVD), 123
Missed visual cues (railway), 91
Mississippi Valley Association of State Highway Departments, 19
Mobile-office services, as distractional devices, 74
Mobile telephones
cooperative ITS technologies and, 221
provision of traffic information services via, 220, 232
Mobility, as goal of driving, 138
Mockups, as discriminative stimulus, 145–146
Modified E Series letterset, 36
vs. Clearview font, 126
Monash University Accident Research Centre, 50, 237
Montoro, Luis, 244, 251–252
Motivational factors, 136–138, 240
defective function of existing sign system attributed to, 156
Movement area guidance signs (MAGS), 98
human factors issues in, 100–104
Multiple-choice tests, in questionnaire-based studies, 157
Multistatic dependent surveillance (MDS) system, 106, 107
Myopia, 117

N

National Codes of Practice for the Defined Interstate Network (Australia), 88

National Highway Traffic Safety Administration (NHTSA), 232
Navigational aids
  enhancing mobility of older drivers, 116
  meeting needs of older drivers, 129
  in VMS systems, 186
Network messages, in VMS systems, 194
New signs, enhancing effectiveness/conspicuity, 56–57
NextMAP project, 218
Ng, Roy, 242, 250
Night driving
  limited by age-related visual decline, 116
  sign illumination enhancing sign detection, 149
Nighttime visibility
  on airport runways and taxiways, 103
  applicability to VMS design, 186
  degrading symbol legibility for older drivers, 123–124
  on roads, 32
No Entry signs, 20
No Hitchhiking symbol, legibility distance of, 123
No Parking signs, comprehension by U.S. and Canadian drivers, 179
No Stopping signs, comprehension by Arab, U.S., and Canadian drivers, 179
Noncompliance-punishment contingencies, 146
Nonstandardization of signage, 72
Nontraffic messages, displayed on VMS systems, 193
Notices to airmen (NOTAMs), 109

O

Oblique astigmatism, 117
Oblique bars, in warning signs, 163–164
Obsolescence, of conventional traffic signs, 235–237
Oceanic drivers, and sign comprehension, 170
Octagonal stop signs, 44
  history of, 19
Ocular movements, age-related decrease in range of, 120
Offline systems (railway), 90
On-board breathalyzer units, 220
On-board tachographs, aiding in speed limit enforcement, 144
Onboard units (OBUs), in vehicle-based technologies, 215
Oncoming vehicle warnings, new pictograms needed for VMS systems, 205
Operating manuals, for in-vehicle technologies, 234
Operational advice, in railway signage, 84
Operational error, causing runway incursions, 104
Operator characteristics, and sign conspicuity, 121
Operator visual abilities, 121
Optical flow, 76
Optical requirements, of signs, 31–33
Orange signs, 45
Organizational issues, associated with airport activity, 100
Out-of-date information
  effects on aging drivers, 120
  on VMS systems, 202

P

Paravisual displays (PVDs), 109
Parking behaviors, real-time information via intelligent speed adaptation (ISA) systems, 217–218
Parking citations, correlated with sign comprehension, 179
Passenger information signs (railway), 90
Passing zones, problematic for older drivers, 229
Pattern recognition, 39
Pavement markings, 26
  better understood by females, 173
Peak periods, VMS message displays only during, 192–193
Pedestrian crossings, 22
  application of VMS to, 186
Pedestrian incursions, on airport runways and taxiways, 104
Pedestrians, need for conventional traffic signs by, 236
Peirce, Charles S., 203
Penalty warning signs, increasing compliance, 144
Perception-response time, age-related problems with, 125–126
Perceptual problems, 6
Peripheral contrast sensitivity, age-related decline in, 119
Peripheral vision, 34
Permanent International Association of Road Congresses (PIARC), 19
Permanent speed boards (railway), 88
Permissive signs, 26
Pfister, Peter, 242, 249
Phonetic alphabet, airport taxiways lettered by, 96
Physical response, to traffic signs, 40
Piano keys, marking thresholds in airport runways and taxiways, 109

Pictograms
development in controlled and nonlocal manner, 210
Finnish use of Congestion pictogram, 202
implications of introducing new, 205–206
Japanese icon orientation in VMS systems, 203
number in VMS systems, 188
perception and comprehension problems with, 205
poor American comprehension in VMS displays, 204
selection and comprehension problems in European VMS systems, 204
Pictorial elements
as fundamental in VMS systems, 194, 195
of signs, 29
in VMS systems, 187, 189
Pilot deviations, causing runway incursions
in Europe, 105
in U.S., 104
Pilot experience, as factor in runway incursions, 105
Pilots
information needs of, 108
poor training in ICAO airport signage standards, 100–101
requiring legibility of runway and taxiway signs, 101
U.S. interventions directed at, 106
Planning functions of driver behavior
use of in-vehicle navigation systems for, 222
VMS role in, 186
Pocket PCs
cooperative ITS technologies using, 221
provision of traffic information services via, 232
use by emerging IVNS technologies, 222
Police presence, strengthening driver response to speed limit signs, 144
Port lights, 22
Potter, Robert, 242, 249–250
Precision, of traffic signs, 39
Preflight checklists, 103–104
Presbyopia, 117
Prioritization, of redundant messages from multiple sources, 231
Priority signs, 3
Procedural guidance, 234
Prohibitory signs, 3, 26
Prometheus traffic management system, 200
Prosymbolic/proiconic systems, in VMS displays, 203–204
Protocol on Traffic Signs and Signals, 21
Psychology of motivation, 136–147
Public route numbering, 18
Punishment, 136

Q

Queensland Rail, aspect sequencing in, 85
Questionnaire-based studies, limitations and biases of, 157

R

RACS traffic management system (Japan), 200
Rail influences, on history of signing, 21–22
Railway collisions, leading to improved signage, 83
Railway operations, 13
Railway signage, 83–93, 242
application of VMS to, 186
future of, 92–93
operational advice via, 84
safety-critical information on, 84
signals passed at danger (SPAD), 91–92
taxonomy of safety-critical signals, 84
advisory/information signs, 90
colored-light signals, 85–86
fixed board signals, 86–88
mechanical semaphore signals, 84–85
passenger information signs (static/dynamic), 90
speed boards, 88–89
Railways, as first form of mechanized transport, 83
Readiness to respond, enhanced by warning signs, 139
Reading, age-related difficulty in, 116
Reading schemas, in VMS systems, 194
Real-time graphic displays
based on real-time traffic conditions, 214
in dynamic intelligent speed adaptation (ISA) systems, 217
in European VMS systems, 204
in future multiple-source traffic information systems, 226
Japanese leadership in VMS systems, 204
railway passenger information signs, 90
Rear-view mirrors
size affecting sign legibility, 129
transient glare from, 117
Reassurance direction signs, 45
Recall method, 52
repeating signs enhancing, 145
of signs "just passed," 135
Recognition distance

effects of driver age on, 122
for symbolic *vs.* text-based signs, 10
Recommended practices, for airport signing, 98–100
Rectangular signs, 44
for information signs, 7
Red color
as background color in mandatory instruction airport signs, 98
maritime symbolism of, 22
Red Flag Act, 22
Red signs, 45
Redundancy of information, 54, 73
concerns in human-machine interface design, 231
eliminating in VMS systems, 190
reducing sign clutter by avoiding, 151
in traffic signs, 148
Refractive problems, age-related, 117
Regan, Michael, 245, 252–253
Regulatory compliance, as high driving priority, 76
Regulatory signs, 2–3, 214
basic message format for VMS systems, 194–195
better understood by females, 173
comprehension by Arab, Asian, and Western drivers, 171
inside vehicles, 216–218
intelligent speed adaptation (ISA) technology, 217
in Variable Message Signs, 185
Reinforcement, 136
Relative intensity, 149
of VMS messages, 227
Reliable contingencies
Children Crossing sign with flashing lights, 143
excessive speed and police-enforced penalty, 144
Reminder signs, increasing compliance, 145
Remote service provider, links from in-vehicle navigation systems, 222
Repeating signs, as determinant of driver awareness/sign conspicuity, 57–59
Repetition priming effect, 58–59
Replacement time, based on retroreflectivity, 28
Rerouting, via VMS systems, 187
Research approaches, 79, 157–159
Response time
age-related problems with, 125–126
considerations in human-machine interface design, 231
and sign legibility, 121
Restricted lane warnings, new pictograms needed for VMS systems, 205
Restrictive signs, 3
Retinal illumination, age-related decline in, 117, 119
Retroreflective sheet, 27–28
enhancing sign detection at night, 149
Reverse collision warnings, 220
Right of way signs, historical development of, 20
Risk management strategies, to reduce railway signs passed at danger (SPAD), 92
Risk-taking
in sign compliance, 6
vs. sign comprehension in accident involvement, 177–179
Road conditions
adjusting speed limits to via VMS, 202–203
in-vehicle guidance regarding current, 223
transmitted via dedicated car radio frequencies, 219
Road designers, responsibilities in human-machine interface design, 231
Road infrastructure expansion, limits to, 200
Road maintenance costs, 200
Road Narrows symbol, comprehensibility by age, 125
Road sensors, 214, 216
Road sign echoing, 74
Road Work Ahead sign, 142
iconic representation in VMS systems, 204
as unreliable contingency, 141
Roadside sensors, in infrastructure-based ITS technologies, 215
Roadway curvature, problematic for older drivers, 229
Rotating advertising space, 77
Route numbering, 18
history of U.S., 18
Route signs
high priority for attention, 149
history of, 17–18
Routine approach to road sign interpretation, 204
Rows, number in VMS systems, 188, 190
Runway and taxiway markings
enhancing for poor visibility conditions, 108
as factor in runway incursions, 105
Runway hold position signs, 98
institution of laser lights at, 106
research to improve visibility of, 106–107
Runway incursions, 104–105
automated air traffic control alerts for, 107
case study from Singapore, 108–111
interventions in prevention of, 105–108
runway hold signs ignored during, 106
Runway numbers, 98

Runway Safety Blueprint 2002-2004, 106
Runway safety programs, 106
Runway warning alert system, 106, 107
Runways, 242
  development of, 95–96
  naming conventions for, 96

S

Saccadic eye movements, 119–120
Safe behavior
  as goal of warning signs, 53
  unrelated to sign comprehension, 177–179
Safety
  in aviation, 242
  compromised by complex in-vehicle technologies, 234
  as function of Intelligent Transport Systems, 185
  as goal of driving, 138
  as highest driving priority, 76
  of signs/poles, 46
  undermined by proliferating technological innovation in VMS, 209
Safety benefits
  of Intelligent Transport Systems (ITS), 214
  of Variable Message Signs (VMS), 226
Safety characteristics, correlation with sign comprehension, 179–181
Safety-critical information, in railway signage, 84
Safety vehicle using adaptive interface technology (SAVE-IT), 232
Salience, of VMS messages, 202
Same-different categorization task, 59
Satellite monitoring, speed capture using, 14
Saudi Arabia, correlation of sign comprehension wtih age, 174
Search conspicuity, 33
  and sign borders, 62
Search strategy, and sign conspicuity, 121
Seat belt usage
  automated reminders or seat belt interlock systems, 220
  correlation with sign comprehension, 179
Semantic priming effect, 58
Semaphores
  mechanical railway signals, 84–85
  use in maritime signing, 22
Semiotics, VMS implementation of, 203–204
Series circuits, illuminating airport runway and taxiway signs, 102
Series E font. *See* Modified E Series letterset
Shapes, of signs, 7
  international differences for warning signs, 163
Sharpness, as determinant of conspicuity, 33
Shield-shaped signs, 44
Sign angles, in airport runways and taxiways, 100
Sign borders, enhancing conspicuity, 62–63
Sign color, 42–44
Sign comprehension, 4. *See also* Comprehension
  differences among Arab, Asian, and Western drivers, 170–171
Sign conspicuity, 53
Sign detection, 148–151
  adversely affected by age-related visual decline, 116
  applicability to VMS, 186
  enhanced by illuminated signs at night, 149
  prior to recognition, 141
Sign deterioration
  and aging operators, 120
  and visual ability to resolve text-based signs, 124
  in VMS systems, 206
Sign effectiveness, undermined by unreliable contingencies, 141–145
Sign fonts, legibility variations in, 126
Sign irrelevance, with unreliable contingencies, 141–142
Sign location
  external to vehicle in conventional traffic signs, 214
  on-board in intelligent transport systems, 214–215
Sign luminance levels, 28, 32
Sign materials, history of, 21
Sign naming, 50, 53
Sign pictures. *See also* Pictograms; Symbolic/pictorial signs
  in VMS displays, 203
Sign placement issues, 31
attention value and, 149
poor placement effects on aging drivers, 120
Sign reading, perspectives for VMS systems, 204–206
Sign recognition, correlated with driving experience, 141
Sign registration, enhanced by repeated presentation of signs, 59
Sign safety, 45
Sign shapes, 42–43
  common meanings of, 44
Signage design
  based on information-processing capabilities of users, 80
  meeting human needs, 77–78
  strategies for older operators, 128

strategies for traffic professionals, 128–129
Signal nameplates, railway, 90
Signal repeaters, railway, 86
Signals passed at danger (SPAD), 91–92
human factors in prevention of, 92
Signposts, historically funded by public subscription, 18
Simons, Mathew, 17
Simulator systems, use in optimizing signage design, 130
Simultaneous presentation of signs, 60–61
Singapore Ministry of Transportation accident analysis, 109
Single sign, faster response to, 61
Sirens, and dynamic variation in intensity, 150
Situated approach to road sign interpretation, 205
Situational awareness (SA), 78, 79
airport runway and taxiway issues, 103–104
pushing pilots and air traffic controllers to limits of capacity, 112
in railway systems, 92
reasons for occurrence in runway incursions, 105
Size
applicability to VMS design, 186
as determinant of conspicuity, 33, 55, 121
specifications for airport signing, 101–102
Slippery roads, real-time information via intelligent speed adapation (ISA) systems, 217
Slippery When Wet signs
comprehension in Arab region *vs.* U.S., 170
pictogram problems in VMS systems, 207
Small print, effects on aging drivers, 120
Smooth pursuit eye movements, 119–120
Smooth pursuit gain, 119
SNCF (France), automatic train protection system, 93
Snellen letter acuity, 122
Snow banks, obscuring airport runway and taxiway signs, 103
Snow warnings, new pictograms needed for VMS systems, 205
Socioeconomic characteristics, correlation with sign comprehension, 179–181
Spacing, of letters, 36
Spain, sample warning sign, 5
Spanish Road Directorate, 197, 210
Spatial and temporal contrast, age-related issues with, 73, 118–119
Spatial vision, age-related decline in, 118
Special regulation signs, 3
Specific intensity per area (SIA), 27, 28
Speed boards (railway), 88–89
Speed capture, using satellite monitoring, 144
Speed limit signs
adjusting speed limits to road conditions in VMS, 202–203
advisory or mandatory on bends and curves, 147
compliance improvements with VMS technologies, 203
failing to control behavior, 142
in GPS in-vehicle systems, 151
ignored with impunity, 144
in-vehicle in global positioning systems, 151
repeating at intervals, 145
text-based in VMS systems, 204
vehicle-activated signs as interactive, 215–216
in VMS systems, 187
Speed of information exposure, 76
Speed signaling (railway), 85
Speeding citations
and risk-taking behaviors, 179
unrelated to sign comprehension, 179
Stability, of posted signs *vs.* VMS, 207
Stack signs, 45
Standard flight instruments, replacing with LCD screens, 108
Standardization, 4, 240
of airport signing, 98–100
of design guidelines for human-machine interfaces, 230
European *vs.* American traffic sign symbols, 21
internationalization and use of icons on road signs, 203
in updating digital maps, 218
vs. innovation in VMS coding schemes, 227
vs. proliferation of local/national VMS pictograms, 207
Starboard lights, 22
Static acuity, age-related decline in, 118
Static information
in conventional traffic signs, 214
vs. real-time information technologies, 214–223
Statistical regression, 244
of sign comprehension studies, 179–181
Statutory placement requirements, 72
Stimulus intensity, dynamic variation in, 149–150
Stimulus-response patterns
issues in future design of VMS systems, 227
issues in human-machine interface design, 233
STOP boards, 86
Stop signs
as example of discriminatory stimulus, 140
history of, 19
Stopping distances, 43, 44

Strategic actions, of VMS, 187
Stroke width, of sign characters, 35
Study biases, questionnaire-based studies, 157
Suitability for purpose, 77
Surface incidents, *vs.* runway incursions, 104
Swedish-Finnish bilingual messages, 192
Swiss Federal Railways, automatic train protection system, 93
Sydney Airport aerodrome chart, 97
Symbolic/pictorial signs. *See also* Icons; Pictograms
- advantages of, 29–30
- driver knowledge declining with age, 175
- higher error rate of detection than text-based signs, 157
- icon orientation of European VMS systems, 203
- identified more quickly and accurately, 60
- optimizing legibility of, 126–127
- perception-response time for, 125
- recommendations for VMS systems, 193
- superiority over text, 10
- *vs.* icons in VMS systems, 203
- *vs.* text-based signs, 9–11
  - legibility for older drivers, 123–124

Symbols, 29
- complexity affecting legibility distance, 121
- incomprehensible/illegible, effects on aging drivers, 120
- use of large in signage design, 129

T

Tachistoscopic presentation, of road signage, 79
Tactical actions, of VMS, 187
Tactile information, advantages of, 228
Tarmac areas, surface incidents on, 104
Tawiway location signs, 99
Taxiway guidance signs, 99
Taxiways, naming conventions for, 96
Telematics waves, 200
Temporal summation, protracted in older drivers, 119
Temporary speed restriction boards (railway), 88, 89
Text-based signs, 72
- combined with pictograms in VMS systems, 193–196, 204
- format order for VMS messages, 191
- length and layout affecting legibility of, 121
- lower error rate of detection than for symbolic signs, 157
- optimizing legibility of, 126
- poor comprehensibility of selected, 125
- text orientation of U.S. VMS systems, 203
- in VMS systems, 187, 189
- *vs.* symbolic/pictorial signs, 9–11

Three-aspect signals, 86
Throats, as traffic calming measures, 145
Timely action, 40
- questionable for AMASS automated alert system, 107

Timing, of multiple-message presentation, 231
Toll roads, 18
Tornay, Francisco, 241, 248
Tort liability, agencies held immune from, 164
Tourist information, provided by on-board ITS systems, 221
Towable barriers, on airport runways, 110
Traffic accidents. *See* Accidents
Traffic alert collision avoidance systems (TCAS), 108
Traffic and Road Safety University Institute-INTRAS (University of Valencia), 197, 210
Traffic calming measures, 145, 146
Traffic control devices
- role of, 26–27
- sign definitions, 27
- sign materials, 27–29

Traffic density, 200
Traffic flow improvement, as function of Intelligent Transport Systems, 185
Traffic information
- accessibility to all drivers, 226
- advantages of real-time, 226, 228
- multiple future sources of, 225–226
  - and obsolescence of conventional traffic signs, 235–237
  - road signs, VMS, and in-vehicle displays, 227–230
  - road signs plus VMS, 226–227

Traffic information systems, 219–220
combining with in-vehicle navigation systems (IVNS), 222
Traffic islands, 36
Traffic lights, technology similarity to in-ground inductance loops, 107
Traffic management
- Prometheus and Drive systems, 200
- *vs.* traffic control, 200

Traffic message channels, 219
Traffic sign definitions, 27
Traffic sign materials, 27–29
Traffic sign requirements, 42–46
Traffic sign theory, 29–42
Traffic signals, 26
- similarity of VMS properties to, 226
- *vs.* traffic signs, 27

Traffic signs, 26. *See also* Transport signs
  as additional discriminative stimuli, 148
  characteristics of conventional, 214
  combined with variable message signs (VMS), 226–227
  as communications system, 73–74
  constraints on
    changing information needs, 75–76
    environment and infrastructure, 75
    in-vehicle echoing of external road signs, 76
    information selection, 76
    visual demands of journey, 76–77
  conventional static *vs.* intelligent transport systems, 213–214
  costs and benefits of, 72–73
  defined, 2
  as discriminative stimuli, 138–140
  future obsolescence of, 235–237
  future pedestrian needs for, 236
  as future source of traffic information, 225
  inside vehicles
    advanced driver assistance systems, 220–221
    guide information, 221–223
    regulatory information provided by, 216–218
    traffic information systems, 219–220
    warning information provided by, 218–219
  limitations of, 72
  pervasiveness of, 135
  posted *vs.* VMS, 206–209
  recall method of research, 52
  risks of, 26
  studies and techniques in evaluating, 158–159
  vs. traffic signals, 27
Traffic states, VMS categories for describing, 189
Train departure times, 90
Train separation, as goal of railway signage, 83
Training, in use of in-vehicle technologies, 235
Trangular yield sign, historical development of, 20
Transponder signals, dependence of automated alert systems on, 107
Transport signing research, 6–12
Transport signs
  age-related problems with, 116–130
  defined, 2
  effectiveness of, 49–50
    measuring, 50–53
  history of, 13
  requirements of good, 4–6
Travel efficiency, as high driving pariority, 76
Travel information and communication systems (TICS), 74
Travel speed, affecting message length in VMS systems, 190
Triangular signs, for warning messages, 7, 163
TVM90 system (France), 92
Two-aspect signal repeater, 86
Two-aspect signals, 86
Two-frame messages, in VMS systems, 194

U

U.K. Driving Standards Agency, 135
Uncertainty, using signs to reduce, 146–147
Unexpected vehicles
  age-related problems in dealing with, 116
  as cause of accidents, 177
Unfamiliar routes
  avoidance in aging drivers, 116
  using signs to reduce uncertainty of, 146–147
United Nations Economic Commission for Europe, 2, 161
Units, number in VMS systems, 188, 190
University of Michigan Transportation Research Institute, 50
Unreliable contingencies
  undermining sign effectiveness, 141–145
  in VMS systems, 202
U.S. Federal Highway Administration, 3
Useful field of view (UFOV), age-related decline in, 119
User acceptability
  and design of future VMS systems, 227
  of human-machine interface technologies, 233–234
User characteristics, determining signage effectiveness, 74

V

Variable intelligent speed adaptation systems, 217
Variable Message Signs (VMS), 4, 11, 12, 13, 29, 241, 244
  ability to serve multiple road functoins, 185
  administrative problems in implementing changes, 207
  advantages and disadvantages of, 185–186
  balancing innovation with standardization, 206–207
  bilingual messages in, 191
  blank *vs.* generic messages, 202
  causing unnecessary distractions to motorists, 193

characters per row, 188–189
competing with other driving priorities, 197
design considerations, 187–188
design recommendations for better use of, 186
device heterogeneity of, 209
display of unimportant information, 193
display parameters, 188
documented safety benefits of, 226
driver trust issues, 72
as emerging technology, 214, 215–216
European use of pictograms with text, 189
expected improvements in, 201–210
extending concepts of, 209–210
flashing lights and supplementary features of, 189
flexibility of, 185
format order for text messages in, 191
as future source of traffic information, 235
general informational requirements of, 189
generic messages *vs.* blank, 202
historical development of, 200
in-vehicle issues, 76
interactive vehicle-activated signs, 215–216
interpreting message absence in, 192–193
issues and disadvantages of, 226–227
lack of technological standards in, 209
lane closure message formats, 191
local or national proliferation of new pictograms in, 207–209
message content in, 189
message formats, 190–191
message length, 190
multiple presentation system problems in, 206
nontraffic message design form, 193
number of pictograms in, 188
number of rows in, 188, 190
potential advantages of, 226
problems of multiple content potentials, 206
prosymbolic/proiconic systems in, 203–204
redesigned daily by multiple sign producers, 206
rerouting messages in, 194
technical display capability problems with, 206
text messages on, 189
two-frame or alternating messages, 194
use of Congestion pictogram in Finland, 202
visual elements of, 187
Vehicle-based ITS technologies, 215
Vehicle braking response time, 41
Vehicle designers, responsibilities in human-machine interface design, 231
Vehicle incursions, on airport runways and taxiways, 104
Vehicle speed, age-related problems judging and dealing with, 116
Vehicle stopping distances, 41
Verbal protocols, 79
Vienna Convention on Road Signs and Signals, 157, 161–162
amendments to, 162, 164
applicability to VMS systems, 207
and VMS message format rules, 194–195
Vigilance, enhanced by warning signs, 139
Visibility
of airport runway and taxiway signs to pilots, 100
decreased in older drivers with low illumination, 116
determined by background and brightness contrast within a sign, 149
enhancing airport runway markers with respect to, 108
requiring visual search behavior, 51
size and, 60
of traffic signs, 30
Visibility distance, for symbolic *vs.* text-based signs, 9–10
Visibility problems, 6
Visual acuity, 6, 243
advantages of augmented signage technologies for, 228
age-related problems with, 73, 116, 118
Visual aids, for airport navigation, 98–100
Visual attentional demand, 74
Visual behavior, measuring, 51
Visual clutter
See also Clutter
reducing, 63–65
Visual complexity, affecting legibility distance, 121
Visual decline. *See* Age-related visual decline
Visual demands on drivers, 71
Visual field, reduction with age, 116
Visual information acquisition strategies, 79
Visual occlusion, 78
Visual overload in signage, 76
Visual search, 51
age-related decline in efficiency of, 120
enhanced for newer signs, 57
Voice warnings, enhancing compliance, 151

W

Warning information
in advanced driver assistance systems, 220–221
emerging technologies providing, 218–219

in intelligent speed adaptation (ISA) systems, 217
lack of timeliness in AMASS alert system, 107
Static *vs.* real-time, 219
in traffic information systems, 219–220
Warning lights, increasing sign conspicuity, 53
Warning signs, 18–19, 214
basic message format for VMS systems, 194
behavioral functions of, 139
behavioral testing of, 7
comprehension by Arab, Asian, and Western drivers, 171
diamond *vs.* triangular shapes of, 163
drivers over 60 ignoring, 175
as examples of discriminative stimuli, 138–140
identified more quickly if symbolic, 60
legibility distance of, 123
more effective than indication signs, 53
number of times viewed, 51
in Variable Message Signs, 185
vs. indication signs, 7–8
Weather conditions
advantages of augmented signage technologies under adverse, 228
obscuring airport runway and taxiway signage, 102
obscuring road signage, 77
real-time information via intelligent speed adaptation (ISA) systems, 217
Web browsing
in-vehicle, 74
for information about real-time traffic conditions, 220
Western drivers, most knowledgeable in sign comprehension, 170
Whole words, detecting, 36–37
Wind directions, and runway organization, 96
Window of exposure
speed of approach determining, 148
and time limitations on viewing airport signs, 103
Windshield problems, age-related, 116
Wireless application protocol (WAP) services, 220
With-the-rule astigmatism, 117
Word choice in VMS systems, 189
Work zone signs, poorly comprehended in U.S., 170
Workload issues
in airport taxiing and takeoff, 103–104
with constant VMS message display, 193
net effect of augmented signage on drivers, 231
Workload managers, 232
World War II, role of signs during, 72

X

Xenon headlights, 78

Y

YARD LIMIT boards, 86
Yellow signs, 45
Yield signs, historical development of, 20